Current Human Cell Research and Applications

This series covers basic and clinical research on human cells, including molecular diagnostics/targeted therapy, cell therapy, cancer stem cells, regenerative medicine, etc., and provides an up-to-date review of human-cell research. All volumes are contributed by leading experts in the field, and offer valuable resources for both cell biologists and clinical researchers in the areas of oncology, stem cell biology, regenerative medicine, and clinical medicine including gynecology, gastroenterology, etc.

Current Human Cell Research and Applications will be published in partnership with the Japan Human Cell Society

More information about this series at http://www.springer.com/series/15107

Haruhisa Inoue • Yukio Nakamura
Editors

Medical Applications of iPS Cells

Innovation in Medical Sciences

Springer

Editors
Haruhisa Inoue
CiRA, Kyoto University
Kyoto
Japan

iPSC-based Drug Discovery
and Development Team
RIKEN BioResource Research Center
Kyoto
Japan

Medical-risk Avoidance based
on iPS Cells Team
RIKEN Center for Advanced
Intelligence Project
Kyoto
Japan

Yukio Nakamura
Cell Engineering Division
RIKEN BioResource Research Center
Tsukuba-shi
Ibaraki
Japan

ISSN 2522-073X ISSN 2522-0748 (electronic)
Current Human Cell Research and Applications
ISBN 978-981-13-3671-3 ISBN 978-981-13-3672-0 (eBook)
https://doi.org/10.1007/978-981-13-3672-0

Library of Congress Control Number: 2019930482

This Springer imprint is published by the registered company Springer Nature Singapore Pte Ltd.
The registered company address is: 152 Beach Road, #21-01/04 Gateway East, Singapore 189721, Singapore

Preface

The creation of chimeric animals and the discovery of embryonic stem cells have had a profound effect on the study and treatment of diseases. The resulting animal models have enabled extraordinary insights on the very beginnings of a disease that can only be identified in its late stages in human patients. Of course, these models have limitations, mostly being that animals and humans, while sharing a great deal of similarities, differ in significant ways that are best summarized by Judah Folkman's remark, "if you have cancer and you are mouse, we can take good care of you."

The discovery of induced pluripotent stem (iPS) cells, specifically human iPS cells, began a new generation of modeling that invited the opportunity to watch a patient's cells develop the disease phenotype in real time. Understanding how to control the pluripotency network of a somatic cell has enabled scientists to reprogram easily accessible cells into those afflicted by the pathology and observing how development differs in healthy and diseases groups and even within diseased subgroups. That these experiments can be done using human cells invites comparisons with animal models and explanation as to why patients diagnosed with the same disease respond differently to an experimental medicine.

Thus, the very first report of human iPS cells was followed by a surge not only in research that explored the reprogramming mechanism but also in research that sought to exploit the power of reprogramming technology for new medical applications. These applications include the study of the pathogenesis of diseases along with new drug discoveries and cell therapies. In more recent years, the development of complementary technologies, such as gene editing and organoids, has only added to the expectation that iPS cells will provide solutions to what are currently intractable diseases.

Accordingly, we have invited authors who view iPS cells as a tool that advances medical treatments and use them to study disease development and generate cells for transplantation or to find candidate compounds. We have also invited authors who use iPS cells to test their new technologies in gene editing, nanomachines, and organoids. Overall, the intention of this book is to provide readers with an impression of the reach iPS cells have in medicine and how their own research can benefit by incorporating iPS cell technology.

Kyoto, Japan Haruhisa Inoue
Tsukuba-shi, Ibaraki, Japan Yukio Nakamura

Acknowledgments

We want to thank Peter Karagiannis (Center for iPS Cell Research and Application) for his assistance with the editing and writing of this book.

Contents

Part I
Introduction

Chapter 1
Clinical Potential of Induced Pluripotent Stem Cells

Peter Karagiannis

Abstract The ability to reprogram cells into induced pluripotent stem cells (iPSCs) has given a new perspective on cellular identity and cellular development. Since the original report of iPSCs, the range of methods and species in which iPSCs have been achieved demonstrates a universality of the pluripotency network and its maintenance. From a clinical perspective, iPSCs provide a new human cell model to study disease and innovate therapies. The reprogramming of patient cells provides a unique human cell model to investigate pathogenesis. iPSCs have especially strong appeal for the study of rare diseases or diseases that are normally diagnosed at late stage. Experimental therapies and drugs based on iPSC research are now at clinical stage. Additionally, drug repositioning using iPSC models has attracted heavy investment from industry. Despite their potential, iPSCs have inconsistent epigenetics with embryonic cells, which has retarded research on some cell lineages and related diseases. Regardless, there is anticipation that a large number of diseases will be treatable using iPSC-based products in the next decade.

Keywords Induced pluripotent stem cells · Disease models · Cell therapy · Drug discovery · Drug repositioning

The creation of induced pluripotent stem cells (iPSCs) has reshaped the scientific view on cell identity. Somatic cell nuclear transfer (SCNT) demonstrated that a somatic cell has the potential to be reprogrammed into any cell lineage, proving that all cell identities can be disrupted through epigenetic rewriting [1]. iPSCs revealed which factors can initiate the pluripotency network from which all adult cells are derived. The first cells to be successfully reprogrammed into mouse or human iPSCs were fibroblasts [2, 3]. Using a retrovirus, the transcription factors Oct4, Sox2, Klf4, and c-Myc (OSKM) were transiently expressed in the fibroblasts to activate the reprogramming apparatus. The induction of OSKM initiated a series of changes

P. Karagiannis (✉)
Center for iPS Cell Research and Application (CiRA), Kyoto University, Kyoto, Japan
e-mail: peter@cira.kyoto-u.ac.jp

H. Inoue, Y. Nakamura (eds.), *Medical Applications of iPS Cells*, Current Human Cell Research and Applications, https://doi.org/10.1007/978-981-13-3672-0_1

including metabolic, morphological, transcriptional and epigenetic, that cause the fibroblasts to function like embryonic cells. Other combinations of reprogramming factors were shown to successfully reprogram cells to iPSCs around the same time [4]. Since then, proteins, RNA, and even cocktails of only small chemical compounds have been shown capable of reprogramming cells to iPSCs [5]. Similarly, a number of integrative and non-integrative methods have been developed, with each having different performance levels in terms of reprogramming efficiency and clinical safety. Finally, the number of cell types and species that have been reprogrammed has convincingly shown that the reprogramming mechanism is common in all mammals [6, 7].

iPSCs represent a comparable model to embryonic stem cells (ESCs) for the study of cell development. Because they can be generated from patients, in some ways they surpass ESCs for the study of disease pathogenesis; accessible patient cells can be reprogrammed and differentiated into the afflicted cells, allowing researchers to observe in real time the cell develops the pathological state in vitro. The first reported patient-derived iPSCs came from reprogramming fibroblasts of a patient suffering from amyotrophic lateral sclerosis (ALS) [8]. This study showed the iPSCs could be differentiated into motor neurons but did not confirm that the neurons expressed the disease phenotype. This next step was demonstrated a year later for two other neurodegenerative diseases, spinal muscular atrophy and familial dysautonomia [9, 10]. Advances in genome editing have enhanced the use of iPSCs for the study of diseases with clear genetic causes such as muscular dystrophy [11]. They are also resourceful for the study of complex diseases. Using iPSCs, the Muotri Lab found that among these variants, one that disrupts the *TRPC6* gene, which had not been previously implicated in autism, results in altered neuronal development and function [12]. Later work by the lab found a common molecular mechanism in a cohort of patients that could make a promising drug target [13]. Some studies have indicated the degenerating cell is not the best target for intervention. A study that differentiated neural progenitor cells from ALS patients into astrocytes and motor neurons showed astrocytes are toxic to the motor neurons [14]. Although patient iPSCs were not used in that study, such cells provide a plentiful source of astrocytes, neurons, and other cell types including those outside neural progenitor cell lineage for diseases caused by cell nonautonomous factors. The production of these multiple cell lineages from iPSCs in combination with 3D organoid technology, a topic to which another chapter of this book is devoted, has promise to capture the influence of biochemical and biomechanical signaling of the microenvironment on the complex cellular networks responsible for many diseases [15].

1.1 Clinical Applications of iPSCs

Considering that iPSCs function as ESCs and the number of ESC-based therapies that have reached clinical trials, there is anticipation that iPSC-based therapies will cover if not surpass the same range of diseases [16]. The first peer-reviewed paper about a cell transplantation that used an iPSC product was published in 2017. Fibroblasts from a patient suffering from age-related macular degeneration (AMD) were reprogrammed into iPSCs, which were then differentiated into retinal epithelial cell (RPE) sheets and transplanted onto one eye of the patient [17]. AMD treatment makes an ideal prototype, because the eyes are relatively easy to diagnose for side effects (including the risk of tumors, a constant concern with any iPSC-based therapy) and because the number of cells required for the therapy is magnitudes less than for other organs. Details of the study are discussed in Chap. 8. Here, I want to give consideration to the implications for future iPSC-based applications.

Clinical trials that use either ESC products or autologous transplants have been reported for AMD [18]. Autologous iPSCs offer the benefits of both, including the ability to expand the cells indefinitely and reduce the need for immunosuppresants. However, the expectation that autologous iPSC products will become standard for regenerative medicine is naïve, a point even the AMD iPSC study asserts. The AMD operation was successfully done on one patient, but a planned operation on a second patient was canceled because of mutations found in that patient's autologous iPSCs. There was no evidence that any of these mutations were tumorigenic. Nevertheless, recognizing that any failures in first-generation iPSC-based therapies could have severe repercussions on future clinical translation, the researchers decided to use standard anti-VEGF therapy for the second patient. There remains an argument on whether the cancelation was overly prudent, asserting a need for standard criteria to evaluate the safety of iPSC products [19, 20].

Furthermore, even safety had not been an issue, the cost and time of an autologous iPSC-based therapy are untenable. For the AMD patient that received the autologous transplant, it took approximately 1 year for the surgery when considering the time of the fibroblast biopsy to RPE transplantation and at a cost of approximately (US) $1 million. For iPSC-based therapies to serve a wide patient population, allogeneic iPSCs are considered a viable alternative. Preclinical studies using non-human primate models have shown that HLA-matched allogeneic grafts in the heart and brain perform comparably to allogeneic grafts, as good engraftment was found several months after the transplantation and the need for immunosuppresants was reduced [21, 22]. Yet other organs may require their own conformation. One study found that smooth muscles derived from autologous iPSCs could provoke an immune reaction in mice [23]. The same study found, consistent with the clinical AMD trial, that RPE derived from autologous iPSCs did not trigger an immune response. These results could suggest that immunogenicity depends on the cell type. The uncertainties about the causes of the immune response elicited by an iPSC product reiterate the need for exhaustive evaluation of iPSCs and their derivatives.

Acknowledging the shift to allogeneic sources, nations are developing stocks of clinical grade iPSCs generated from the blood of human leukocyte antigen (HLA) homozygous donors (Fig. 1.1). The challenge for these stocks is serving large populations. It has been estimated that lines covering the 20 most frequent haplotypes would serve 50% of European populations but only 22% of African populations [24]. Moreover, as the number of lines increases linearly, the number of donors that must be screened increases exponentially. Even for Japan, which has a relatively

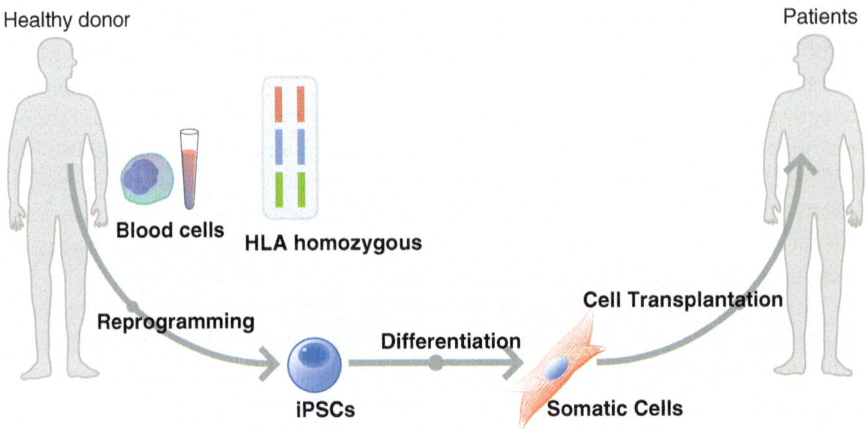

Fig. 1.1 To lower the cost and time it takes for an iPSC-based therapy to reach a patient, organizations are considering stocks that distribute clinical grade allogeneic iPSCs. The iPSCs are prepared from healthy donors with homozygous HLA so as to maximize the number of HLA-matched patients. Stocks will have no direct interaction with the donor or patient. They will be responsible for reprogramming the donor cells into iPSCs and evaluating safety, while third parties will collect donor cells. Cells will be distributed to medical institutes upon request, where they will be differentiated for the purposes of basic research or clinical application

homogeneous population, a bank that stores the 100 most frequent haplotypes, which would cover over 90% of the population, would require recruiting around 100,000 candidate donors [25]. To reduce this burden, organizations developing iPSC stocks are working with the Red Cross, bone marrow banks, cord blood banks, and similar organizations that have extensive donor information to find homozygous donors in an ethically approved manner [26].

In general, iPSC-based therapies are following ESC paradigms for several diseases. In some cases, because ESC therapies are well ahead in clinical development, motivation to transition to iPSC-based therapies can be questioned. On the other hand, some therapies would benefit tremendously by the transition.

Exhaustion describes a state in which T cells can detect cancer peptides but are unable to exert a cytotoxic effect. Studies have shown that iPSC technology can reprogram T cells out of the exhausted state in vitro and into rejuvenated T cells. Rejuvenated T cells behave like memory or effector T cells so that they expand and exert cytotoxic function in response to appropriate cytokines [27, 28]. To retain antigen specificity, only T cells with the appropriate TCR can be used for the reprogramming, because otherwise receptor rearrangement can occur, which would lose the peptide specificity. It has also been shown that the reprogramming of T cells is compatible with the engineering of chimeric antigen receptors, which were recently approved for clinical trials [29]. In principle, with better understanding of lymphopoiesis, it could be possible to rejuvenate exhausted T cells into all stages of T cell development (naïve, memory, effector, etc.) for effective adoptive cellular therapies and to rejuvenate all lymphocytes (natural kills cells, B cells, etc.) [30].

Donors have been providing platelets for platelet transfusion therapy for decades. However, there is growing concern that this dependency is unsustainable. Furthermore, some patients develop refractoriness to platelets following regular therapy, which is why HLA-null platelets have been proposed. Erasure of HLA would remove concern about haplotype matching and permit either ESCs or iPSCs as the platelet source [31]. However, it was iPSC technology that was instrumental in the creation of immortalized megakaryocyte lines (iMKLs), which were made by transiently expressing c-Myc, BMI1, and BCL-XL in iPSCs [32]. Platelets from peripheral blood have only a shelf life of a few days. In contrast, iMKLs can supply platelets for nearly half a year and produce platelets upon request, greatly reducing platelet waste. Megakaryocytes are unipotent progenitors of platelets. Therefore, because the protocol for platelet generation requires fewer steps when the starting source is iMKCLs than iPSCs, the platelet generation should be more efficient, and the risk of contaminants should be reduced. The functionality and safety of the platelets from iPSC-derived sources including iMKLs are satisfactory to the point that the biggest obstacle to clinical translation is the engineering of bioreactors that generate the requisite number of platelets for a single transfusion [33].

Because they can be made from patient cells, the benefits of iPSCs are less in dispute for drug discovery. Prior to iPSCs, patient cells were often of little help to drug discovery because the disease had already reached its late stage. Estimates state that half of the afflicted cells will be lost by the time Parkinson's disease is diagnosed in a patient [34]. In the 10 years since the first report testing chemical compounds on differentiated patient iPSCs, a long list of candidate drugs has been found for neurodegenerative diseases alone [35].

In some cases, the tested candidates were not examples of drug discovery but rather drug repositioning. Drug repositioning is estimated to cut the cost and time of taking a drug to market to one third that for drug discovery [36]. Several examples of drug repositioning from iPSC models have been reported. The first to reach clinical trial is ezogabine, an approved antiepileptic drug that was found to correct deviant Kv7.2/3 potassium channel activity in motor neurons derived from ALS patient iPSCs [37]. The anticancer drug bosutinib, an inhibitor of signaling by two tyrosine kinases, Src and c-Abl, was also found to have positive effects on iPSC-derived neurons from familial and sporadic patients [38]. Because sporadic ALS makes up 90% of ALS cases, which suggests a wide range of molecular causes, it is likely that any experimental drug will only benefit a subset of the patient population. iPSCs provide a cheaper human model compared with animal models to test the drug and also the opportunity to stratify patients into predicted positive responders and predicted negative responders early in the drug development. Fibrodysplasia ossificans progressiva is estimated to inflict one of every two million people worldwide. This disease leads to soft tissue regenerating as bone tissue following trauma or inflammation. The excess bone growth prevents normal functions including respiration. Using patient iPSCs, researchers found that aberrant mTOR signaling contributed to the disease and that rapamycin, an approved drug for immunosuppression, could inhibit the abnormal ossification [39]. A clinical trial based on these findings was approved in Japan. Finally, an iPSC model for thanatophoric dysplasia type I, which causes severely short limbs and underdeveloped organs, led to the discovery that statins can promote bone growth [40]. Patient iPSCs were induced chondrogenically, and the cartilage was transplanted into mice. Cartilage prepared from differentiation protocols that included statin in the culture showed normal growth, whereas cartilage without statin did not. The discoveries of rapamycin and statin using iPSC models have contributed to phenotypic drug screening, which could offer a cheaper approach than targeted drug screenings for complex or rare diseases [41]. Indeed, the statin finding was the catalyst for a massive academic-industry collaboration that aims to be a model for translating basic research [42].

Cost reduction also comes in the form of drug toxicity studies. Many drugs that pass preclinical tests are withdrawn due to cardiac arrhythmias. To test for toxicity, ion channel genes are overexpressed in immortalized cell lines such as cancer cells. iPSCs, on the other hand, enable the investigation of channel responses in patient cells (myocytes in the case of arrhythmias) to a drug [43]. One obstacle to the full adaption of iPSC models for drug toxicity is concern about the maturity of the differentiated cells. Cardiomyocytes derived from iPSCs show fetal properties and thus have different electrophysiological and morphological properties compared with adult cardiomyocytes [44]. Another is recapitulating cellular interactions in the heart. Proper heart function depends on a strict coordination of different cell types. A recent study using iPSCs demonstrated that proper mixing of heterogeneous populations could recapitulate torsade de pointes caused by ion channel blockers in 3D cardiac sheets [45].

1.2 Obstacles to Clinical Application

iPSCs have a number of unique caveats regarding clinical application, foremost being genomic instability. Chromosomal aberration, copy number variation, and single nucleotide variations have all been observed in iPSCs. The reprogramming itself is one possible cause, but so too are the culture duration and preexisting variations in the somatic cells [46]. Indeed, it has been reported that iPSCs prepared from umbilical cord blood show less single nucleotide variations than those prepared from adult fibroblasts [47] and that the frequency of mutations in iPSCs increases with the age of the donor [48]. The above considerations and uncertainties will delay the global use of iPSCs in the clinic, but indicate iPSCs can be used upon proper selection and evaluation. In addition, one of the major excitements about iPSCs as a cell therapy source is the expectation that they will perform equally to ESCs but circumvent the need of embryo destruction. However, iPSCs have their own distinct ethical concerns. In principle, using iPSC technology it is possible to generate clones by reprogramming somatic cells into germ cells [50]. Much like how in vitro fertilization changed our mind-set of what is tolerable for pregnancy, iPSCs have the power to do the same. Additionally, clear consent from donors is required, as suddenly any blood donation has the potential to be used in ways that go well beyond hematological purposes.

There are also general issues with the use of any PSCs, be they iPSCs or ESCs. In normal development, progress from the zygote to the blastocyst involves an erasure of epigenetic memory. In somatic cells too, erasure of epigenetic memory must occur for reprogramming. Both ESCs and iPSCs are expected to recapitulate all stages of development beginning from preimplantation. Therefore, they should faithfully mimic embryonic cells, but there is already evidence that they do not. Pluripotency has been divided into naïve and primed states, with the former defining higher self-renewal capacity and the latter having higher sensitivity to lineage cues. Naïve pluripotent stem cells (PSCs) include mouse ESCs, and primed PSCs include mouse epiblast stem cells (EpiSCs). Unlike mouse PSCs, human ESCs and iPSCs have until recently only shown primed pluripotency. Naïve pluripotent human PSCs have been obtained by resetting primed PSCs in culture [51, 52], but there is increasing evidence that these cells do not recapitulate the true nature of embryonic cells. Collier et al. reported four surface markers specific to the naïve state in PSCs, but one of these, CD7, is not expressed in inner cell mass [53]. Similarly, mouse ESCs show epigenetic aberrations from embryonic cells that put doubt into their pluripotency quality [54, 55]. This deviation could explain why some cell types have proven more difficult to derive from PSCs than others. The epigenetics could also contribute to different differentiation efficiencies between iPSC lines to specific lineages [56, 57].

1.3 Conclusion

The shortening in time between the first SCNT amphibians and mammals [1, 58], first mouse and human ESCs [59, 60], and first mouse and human iPSCs [2, 3] shows Moore's law extends to the invention of new biotechnologies that control pluripotency. We should therefore expect a rapid surge in iPSC-based therapies in the next several years. iPSCs present a cellular model of development that matches ESCs in terms of function but adds a more complex topology in that cell identity is no longer unidirectional. From a clinical perspective, iPSCs match ESCs in terms of regenerative medicine without the same ethical controversies. Unlike ESCs, iPSCs can be made from patient cells, providing a unique model to study disease pathogenesis from the pluripotent state. This feature has also welcomed a new means to study experimental compounds that can convalesce the disease state. While further research is necessary to optimize iPSCs, they are already realizing their promise for new insights on how diseases develop and new means for treatment.

Acknowledgments I thank Masaya Todani, Center for iPS Cell Research and Application (CiRA), Kyoto University (Japan), for the figure illustration.

References

1. Gurdon JB. The developmental capacity of nuclei taken from intestinal epithelium cells of feeding tadpoles. J Embryol Exp Morphol. 1962;10:622–40.
2. Takahashi K, et al. Induction of pluripotent stem cells from adult human fibroblasts by defined factors. Cell. 2007;131:861–72. https://doi.org/10.1016/j.cell.2007.11.019.
3. Takahashi K, Yamanaka S. Induction of pluripotent stem cells from mouse embryonic and adult fibroblast cultures by defined factors. Cell. 2006;126:663–76. https://doi.org/10.1016/j.cell.2006.07.024.
4. Yu J, et al. Induced pluripotent stem cell lines derived from human somatic cells. Science. 2007;318:1917–20. https://doi.org/10.1126/science.1151526.
5. Silva M, et al. Generating iPSCs: translating cell reprogramming science into scalable and robust biomanufacturing strategies. Cell Stem Cell. 2015;16:13–7. https://doi.org/10.1016/j.stem.2014.12.013.
6. Ogorevc J, Orehek S, Dovc P. Cellular reprogramming in farm animals: an overview of iPSC generation in the mammalian farm animal species. J Anim Sci Biotechnol. 2016;7:10. https://doi.org/10.1186/s40104-016-0070-3.
7. Polo JM, et al. Cell type of origin influences the molecular and functional properties of mouse induced pluripotent stem cells. Nat Biotechnol. 2010;28:848–55. https://doi.org/10.1038/nbt.1667.
8. Dimos JT, et al. Induced pluripotent stem cells generated from patients with ALS can be differentiated into motor neurons. Science. 2008;321:1218–21. https://doi.org/10.1126/science.1158799.
9. Ebert AD, et al. Induced pluripotent stem cells from a spinal muscular atrophy patient. Nature. 2009;457:277–80. https://doi.org/10.1038/nature07677.
10. Lee G, et al. Modelling pathogenesis and treatment of familial dysautonomia using patient-specific iPSCs. Nature. 2009;461:402–6. https://doi.org/10.1038/nature08320.
11. Li HL, Gee P, Ishida K, Hotta A. Efficient genomic correction methods in human iPS cells using CRISPR-Cas9 system. Methods. 2016;101:27–35. https://doi.org/10.1016/j.ymeth.2015.10.015.
12. Griesi-Oliveira K, et al. Modeling non-syndromic autism and the impact of TRPC6 disruption in human neurons. Mol Psychiatry. 2015;20:1350–65. https://doi.org/10.1038/mp.2014.141.
13. Marchetto MC, et al. Altered proliferation and networks in neural cells derived from idiopathic autistic individuals. Mol Psychiatry. 2017;22:820–35. https://doi.org/10.1038/mp.2016.95.
14. Haidet-Phillips AM, et al. Astrocytes from familial and sporadic ALS patients are toxic to motor neurons. Nat Biotechnol. 2011;29:824–8. https://doi.org/10.1038/nbt.1957.
15. Quadrato G, Brown J, Arlotta P. The promises and challenges of human brain organoids as models of neuropsychiatric disease. Nat Med. 2016;22:1220–8. https://doi.org/10.1038/nm.4214.
16. Trounson A, DeWitt ND. Pluripotent stem cells progressing to the clinic. Nat Rev Mol Cell Biol. 2016;17:194–200. https://doi.org/10.1038/nrm.2016.10.
17. Mandai M, et al. Autologous induced stem-cell-derived retinal cells for macular degeneration. N Engl J Med. 2017;376:1038–46. https://doi.org/10.1056/NEJMoa1608368.
18. Garcia JM, et al. Stem cell therapy for retinal diseases. World J Stem Cells. 2015;7:160–4. https://doi.org/10.4252/wjsc.v7.i1.160.
19. Daley G, Polar Q. Extremes in the clinical use of stem cells. N Engl J Med. 2017;376:1075–7. https://doi.org/10.1056/NEJMe1701379.

20. Karagiannis P, Eto K. Ten years of induced pluripotency: from basic mechanisms to therapeutic applications. Development. 2016;143:2039–43. https://doi.org/10.1242/dev.138172.
21. Morizane A, et al. MHC matching improves engraftment of iPSC-derived neurons in non-human primates. Nat Commun. 2017;8:385. https://doi.org/10.1038/s41467-017-00926-5.
22. Shiba Y, et al. Allogeneic transplantation of iPS cell-derived cardiomyocytes regenerates primate hearts. Nature. 2016;538:388–91. https://doi.org/10.1038/nature19815.
23. Zhao T, et al. Humanized mice reveal differential immunogenicity of cells derived from autologous induced pluripotent stem cells. Cell Stem Cell. 2015;17:353–9. https://doi.org/10.1016/j.stem.2015.07.021.
24. Gourraud PA, Gilson L, Girard M, Peschanski M. The role of human leukocyte antigen matching in the development of multiethnic "haplobank" of induced pluripotent stem cell lines. Stem Cells. 2012;30:180–6. https://doi.org/10.1002/stem.772.
25. Saito MK, Matsunaga A, Takasu N, Yamanaka S. In: Ilic D, editor. Stem cell banking. New York: Springer; 2014. p. 67–76.
26. Azuma K, Yamanaka S. Recent policies that support clinical application of induced pluripotent stem cell-based regenerative therapies. Regen Ther. 2016;4:36–47.
27. Nishimura T, et al. Generation of rejuvenated antigen-specific T cells by reprogramming to pluripotency and redifferentiation. Cell Stem Cell. 2013;12:114–26. https://doi.org/10.1016/j.stem.2012.11.002.
28. Vizcardo R, et al. Regeneration of human tumor antigen-specific T cells from iPSCs derived from mature CD8(+) T cells. Cell Stem Cell. 2013;12:31–6. https://doi.org/10.1016/j.stem.2012.12.006.
29. Maude SL, et al. The effect of pembrolizumab in combination with CD19-targeted chimeric antigen receptor (CAR) T cells in relapsed acute lymphoblastic leukemia (ALL). J Clin Oncol. 2017;35:103. https://doi.org/10.1200/JCO.2017.35.15_suppl.103.
30. Kitayama S, et al. Cellular adjuvant properties, direct cytotoxicity of re-differentiated Valpha24 invariant NKT-like cells from human induced pluripotent stem cells. Stem Cell Reports. 2016;6:213–27. https://doi.org/10.1016/j.stemcr.2016.01.005.
31. Sugimoto N, Eto K. Platelet production from induced pluripotent stem cells. J Thromb Haemost. 2017;15:1717–27. https://doi.org/10.1111/jth.13736.
32. Nakamura S, et al. Expandable megakaryocyte cell lines enable clinically applicable generation of platelets from human induced pluripotent stem cells. Cell Stem Cell. 2014;14:535–48. https://doi.org/10.1016/j.stem.2014.01.011.
33. Thon JN, Dykstra BJ, Beaulieu LM. Platelet bioreactor: accelerated evolution of design and manufacture. Platelets. 2017;28:472–7. https://doi.org/10.1080/09537104.2016.1265922.
34. Bezard E, et al. Relationship between the appearance of symptoms and the level of nigrostriatal degeneration in a progressive 1-methyl-4-phenyl-1,2,3,6-tetrahydropyridine-lesioned macaque model of Parkinson's disease. J Neurosci. 2001;21:6853–61.
35. Avior Y, Sagi I, Benvenisty N. Pluripotent stem cells in disease modelling and drug discovery. Nat Rev Mol Cell Biol. 2016;17:170–82. https://doi.org/10.1038/nrm.2015.27.
36. Nosengo N. New tricks for old drugs. Nature. 2016;534:314–6.
37. McNeish J, Gardner JP, Wainger BJ, Woolf CJ, Eggan K. From dish to bedside: lessons learned while translating findings from a stem cell model of disease to a clinical trial. Cell Stem Cell. 2015;17:8–10. https://doi.org/10.1016/j.stem.2015.06.013.
38. Imamura K, et al. The Src/c-Abl pathway is a potential therapeutic target in amyotrophic lateral sclerosis. Sci Transl Med. 2017;9:eaaf3962. https://doi.org/10.1126/scitranslmed.aaf3962.
39. Hino K, et al. Activin-A enhances mTOR signaling to promote aberrant chondrogenesis in fibrodysplasia ossificans progressiva. J Clin Invest. 2017;127(9):3339–52. https://doi.org/10.1172/JCI93521.
40. Yamashita A, et al. Statin treatment rescues FGFR3 skeletal dysplasia phenotypes. Nature. 2014;513:507–11. https://doi.org/10.1038/nature13775.
41. Moffat JG, Vincent F, Lee JA, Eder J, Prunotto M. Opportunities and challenges in phenotypic drug discovery: an industry perspective. Nat Rev Drug Discov. 2017;16:531–43. https://doi.org/10.1038/nrd.2017.111.

42. Karagiannis P, Onodera A, Yamanaka S. New models for therapeutic innovation from Japan. EBioMedicine. 2017;18:3–4. https://doi.org/10.1016/j.ebiom.2017.03.042.
43. Nozaki Y, et al. CSAHi study: validation of multi-electrode array systems (MEA60/2100) for prediction of drug-induced proarrhythmia using human iPS cell-derived cardiomyocytes-assessment of inter-facility and cells lot-to-lot-variability. Regul Toxicol Pharmacol. 2016;77:75–86. https://doi.org/10.1016/j.yrtph.2016.02.007.
44. Barbuti A, Benzoni P, Campostrini G, Dell'Era P. Human derived cardiomyocytes: a decade of knowledge after the discovery of induced pluripotent stem cells. Dev Dyn. 2016;245:1145–58. https://doi.org/10.1002/dvdy.24455.
45. Kawatou M, et al. Modelling Torsade de Pointes arrhythmias in vitro in 3D human iPS cell-engineered heart tissue. Nat Commun. 2017;8:1078. https://doi.org/10.1038/s41467-017-01125-y.
46. Yoshihara M, Hayashizaki Y, Murakawa Y. Genomic instability of iPSCs: challenges towards their clinical applications. Stem Cell Rev. 2017;13:7–16. https://doi.org/10.1007/s12015-016-9680-6.
47. Su RJ, et al. Few single nucleotide variations in exomes of human cord blood induced pluripotent stem cells. PLoS One. 2013;8:e59908. https://doi.org/10.1371/journal.pone.0059908.
48. Kang E, et al. Age-related accumulation of somatic mitochondrial DNA mutations in adult-derived human iPSCs. Cell Stem Cell. 2016;18:625–36. https://doi.org/10.1016/j.stem.2016.02.005.
49. Rouhani F, et al. Genetic background drives transcriptional variation in human induced pluripotent stem cells. PLoS Genet. 2014;10:e1004432. https://doi.org/10.1371/journal.pgen.1004432.
50. Zheng YL. Some ethical concerns about human induced pluripotent stem cells. Sci Eng Ethics. 2016;22:1277–84. https://doi.org/10.1007/s11948-015-9693-6.
51. Takashima Y, et al. Resetting transcription factor control circuitry toward ground-state pluripotency in human. Cell. 2014;158:1254–69. https://doi.org/10.1016/j.cell.2014.08.029.
52. Theunissen TW, et al. Systematic identification of culture conditions for induction and maintenance of naive human pluripotency. Cell Stem Cell. 2014;15:471–87. https://doi.org/10.1016/j.stem.2014.07.002.
53. Collier AJ, et al. Comprehensive cell surface protein profiling identifies specific markers of human naive and primed pluripotent states. Cell Stem Cell. 2017;20(6):874–890.e7. https://doi.org/10.1016/j.stem.2017.02.014.
54. Choi J, et al. Prolonged Mek1/2 suppression impairs the developmental potential of embryonic stem cells. Nature. 2017;548:219–23. https://doi.org/10.1038/nature23274.
55. Yagi M, et al. Derivation of ground-state female ES cells maintaining gamete-derived DNA methylation. Nature. 2017;548:224–7. https://doi.org/10.1038/nature23286.
56. Kajiwara M, et al. Donor-dependent variations in hepatic differentiation from human-induced pluripotent stem cells. Proc Natl Acad Sci U S A. 2012;109:12538–43. https://doi.org/10.1073/pnas.1209979109.
57. Kim K, et al. Donor cell type can influence the epigenome and differentiation potential of human induced pluripotent stem cells. Nat Biotechnol. 2011;29:1117–9. https://doi.org/10.1038/nbt.2052.
58. Wilmut I, Schnieke AE, McWhir J, Kind AJ, Campbell KH. Viable offspring derived from fetal and adult mammalian cells. Nature. 1997;385:810–3. https://doi.org/10.1038/385810a0.
59. Evans MJ, Kaufman MH. Establishment in culture of pluripotential cells from mouse embryos. Nature. 1981;292:154–6.
60. Thomson JA, et al. Embryonic stem cell lines derived from human blastocysts. Science. 1998;282:1145–7.

Part II
Disease Modelling

Chapter 2
Disease Modeling of Hematological and Immunological Disorders Using Induced Pluripotent Stem Cells

Megumu K. Saito

Abstract Abnormalities in both mature hematopoietic cells and immature stem/progenitor cells can cause various diseases. In these disorders, because of limited availability of immature hematopoietic progenitor cells, induced pluripotent stem cells (iPSCs) have a great advantage for establishing their disease models. In this chapter, researches using iPSC-based disease models for hematological and immunological disorders are outlined.

Keywords Hematopoietic disorders · Immunological disorders · Induced pluripotent stem cells · Disease models · Hematopoietic differentiation

2.1 Introduction

Blood consists of a blood cell component and a plasma component, circulates throughout the body, and is responsible for various functions. The blood cell component mainly consists of erythrocytes, leukocytes, and platelets. These cells all derive from hematopoietic progenitor cells, which themselves arise from hematopoietic stem cells (HSCs). Abnormalities in both mature hematopoietic cells and immature stem/progenitor cells can cause various diseases. In hematological and immunological disorders, it is possible to obtain patient-derived specimens such as peripheral blood and bone marrow cells. However, the frequent collection of large amounts of blood samples is highly invasive to the patients, and functional evaluation of the cells in vitro is sensitive to the patients' condition, such as the status of the treatment and the cytokine milieu. Moreover, it is difficult to obtain large amounts of HSCs or hematopoietic progenitor cells. In this way, induced pluripotent stem cells (iPSCs) have a great advantage for disease models of hematopoietic and immunological disorders. In this chapter, I will outline research using related iPSC-based disease models (Fig. 2.1).

M. K. Saito (✉)
Department of Clinical Application, CiRA, Kyoto University, Kyoto, Japan
e-mail: msaito-g@cira.kyoto-u.ac.jp

H. Inoue, Y. Nakamura (eds.), *Medical Applications of iPS Cells*, Current Human Cell Research and Applications,
https://doi.org/10.1007/978-981-13-3672-0_2

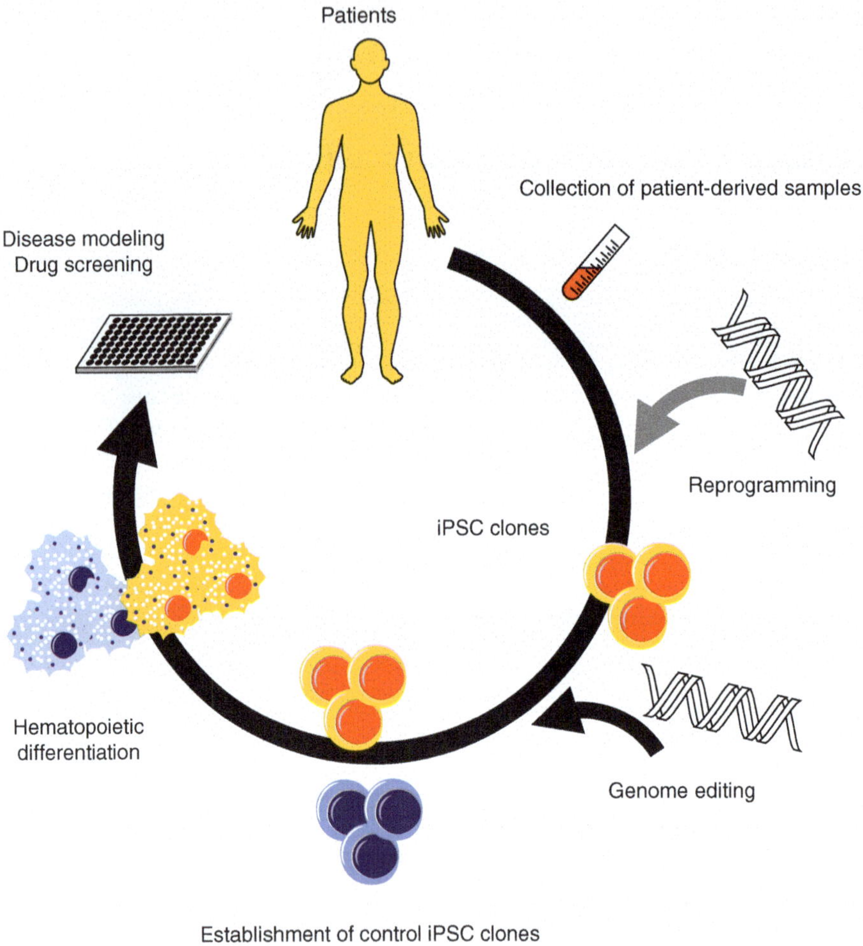

Fig. 2.1 Principles of human iPSC-based hematopoietic disease modeling

2.2 Hematopoietic Differentiation for Disease Modeling

In order to construct iPSC models reflecting the clinical condition of patients, it is necessary to differentiate iPSCs into the cell types responsible for the pathogenesis. Therefore, the construction of a robust differentiation system is essential. Generally, directed in vitro differentiation from pluripotent stem cells (PSCs) is organized to follow in vivo embryogenesis [1–3]. In the case of hematopoietic cells, PSCs differentiate into mature blood cells via primitive streak, lateral plate mesoderm, hemato-endothelial progenitor cells, and hematopoietic progenitor cells. For this purpose, the culture conditions should be switched at each stage.

Normally, the in vitro differentiation of PSCs leads to differentiated progenies that exhibit immature (i.e., not adult) phenotypes. In addition, although PSCs are theoretically able to differentiate into all cell types, there are many cell types for which efficient differentiation systems have not been established, with HSCs being one example. Several recent reports have made progress in this problem, which would advance the establishment of disease models [4–6].

Differentiation systems for hematopoietic cells can be roughly divided into two-dimensional (2D) and three-dimensional (3D) cultures. In 2D cultures, cells are directly adhered to cell culture dishes coated with extracellular matrix [7, 8] or on feeder cells which support the differentiation [9, 10]. In 3D cultures, cells are suspended and then aggregated to form sphere-like structures called embryoid bodies [11, 12]. In embryoid bodies, each cell differentiates according to the culture condition and the cell's 3D positional information. In 2D cultures, hematopoietic cells are easily recovered because they are floating. Furthermore, 2D cultures are easy to scale up. On the other hand, in 3D cultures, cells differentiate and proliferate autonomously by exchanging positional information with each other, making it possible to create complicated organ-like structures (organoids), which may be beneficial for maturing the cells. In other words, each culture system has its own advantages and disadvantages and should be selected based on the purpose of the experiments.

The resulting PSC-derived hematopoietic cells can be used for functional analysis, but the differentiation of PSCs is often time- and labor-intensive. Additionally, the differentiation efficiency and the functional analysis itself can often show high variance. To overcome these issues and stably obtain a large number of mature hematopoietic cells from PSCs, researchers have tried to immortalize PSC-derived hematopoietic progenitors by introducing several genetic factors [13–17]. These immortalized progenitors can give rise to mature and terminally differentiated hematopoietic cells that are well suited for disease modeling (Fig. 2.2). Immortalization provides an unlimited number of functional hematopoietic cells. This feature is especially attractive when using iPSCs derived from patients with hematopoietic and immunological disorders, which can then be used for disease modeling. Indeed, immortalized myeloid cells and multipotent hematopoietic progenitor cells have been used to analyze and identify potential therapeutic compounds [15, 18].

2.3 Disease Modeling of Hematopoietic Disorders with iPSCs

Bone marrow failure is a condition in which hematopoietic output from the bone marrow is insufficient to maintain physiological function of the blood system. In some cases, it is associated with a congenital dysfunction of hematopoietic differentiation. Since iPSC technology can trace hematopoietic differentiation in vitro, it can be used to study related disorders. Fanconi anemia is a congenital bone marrow insufficiency caused by genetic abnormality in the FANC gene family, which codes for a DNA repair enzyme [19–24]. Hematopoietic progenitor cells differentiated from Fanconi anemia patient-derived iPSCs showed impaired hematopoietic colony-forming ability [25, 26]. Drug screenings identified several compounds that

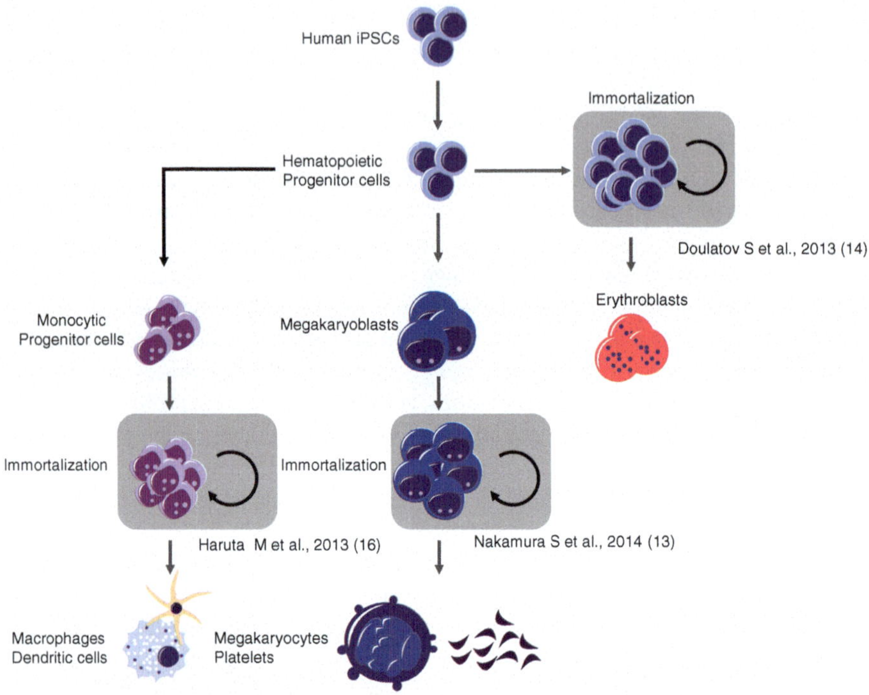

Fig. 2.2 Immortalization of hematopoietic progenitor cells from human PSCs. Hematopoietic progenitor cells can be immortalized by introducing appropriate transcription factors

improved the clonogenecity [27]. In another study, common hemoangiogenic progenitor cells differentiated from Fanconi anemia patient-derived iPSCs showed skewed differentiation toward endothelial lineage and an alteration in the gene expression profile of hemoangiogenic progenitor cells that was associated with the impaired hematopoietic output [28]. Similar screenings using hematopoietic progenitor cells derived from the iPSCs of patients with Diamond-Blackfan anemia, a congenital degenerative erythrocyte differentiation, have been done and showed an autophagy-promoting compound recovers the erythroid differentiation capacity and the expression of globin genes [3]. These data highlight the significance of iPSCs for studying bone marrow failure associated with early hematopoietic defects [28].

Diseases caused by a metabolic dysfunction of hematopoietic cells have been also modeled. Gaucher disease is a typical lysosomal storage disease caused by the loss of glucocerebrosidase enzymatic activity, and it causes anemia and thrombocytopenia [29]. iPSCs derived from Gaucher disease patients showed skewed differentiation to myeloid cells and impaired erythroid differentiation, indicating the abnormal function of hematopoietic progenitor cells in this disease [30].

iPSC models of hematopoietic malignancies are highly useful for studying their pathogenesis [31, 32]. Multiple iPSC clones derived from chronic myelogenous leukemia (CML) blasts have been established [33]. In CML, the formation of chimeric genes due to 9;22 chromosomal translocation is involved in the onset of the disease [34, 35]. When the iPSCs were established into CML blasts, the progeny iPSC clones carried the chromosomal translocation. Interestingly, undifferentiated iPSCs were insensitive to imatinib, a therapeutic agent for CML, but the sensitivity was recovered as the cells were differentiated into hematopoietic cells [32]. Acute myeloid leukemia (AML) blast-derived iPSCs have also been used for disease modeling [36–38]. It was found that AML-iPSCs lost the epigenetic signature of the original somatic cell but reacquired it after hematopoietic differentiation [36], indicating that an iPSC approach can segregate genetic and epigenetic effects during leukemogenesis. Other models of hematopoietic malignancies include an abnormal production of megakaryocytes in familial platelet disorder with predisposition to acute myelogenous leukemia caused by *runt-related transcription factor 1 (RUNX1)* mutations [39], acute lymphoblastic leukemia [40], and myeloid dysfunction in juvenile myelomonocytic leukemia associated with *protein tyrosine phosphatase, non-receptor type 11 (PTPN 11)* mutations [41].

Transient abnormal myelopoiesis (TAM) is a transient hematopoietic disorder affecting about 10–20% of Down's syndrome patients [42]. In TAM patients, abnormal blastic cells transiently proliferate both in peripheral blood and bone marrow blasts in the neonatal period, and 16–30% of TAM patients later develop acute megakaryoblastic leukemia (AMKL). In addition to the original 21 trisomy, TAM blasts carry a somatic *GATA binding protein 1 (GATA1)* mutation, and additional genetic modifications are found in AMKL blasts [43]. Therefore, TAM/AMKL is considered a typical model of multistage carcinogenesis. When iPSCs from TAM blasts were established and differentiated into hematopoietic cells, their potential to commit into erythroid lineage was severely impaired [44]. In line with this finding, an analysis of gene expression profiles of hematopoietic progenitor cells revealed a downregulation of erythroblast-associated differentiation genes, probably due to the loss of GATA1 function. By combining genome editing technology and iPSCs, another group tried to identify the responsible region on chromosome 21 for the onset of TAM [45]. By generating and phenotyping iPSC clones with partial 21 trisomy, a critical 4 Mb region of chromosome 21 that includes the *RUNX1*, *ETS proto-oncogene 2, transcription factor (ETS2)*, and *ETS transcription factor ERG (ERG)* genes was identified [45]. Similar approaches have been conducted for myelodysplastic syndrome (MDS), a preleukemic condition. iPSCs prepared from each stage beginning at MDS progression and ending at leukemia recapitulated the hematopoietic phenotypes at the corresponding stage, and genome editing could resolve the pathological phenotypes [46].

2.4 Disease Modeling of Immunological Disorders with iPSCs

Primary immunodeficiency syndromes (PIDs) are a heterogeneous group of over 130 disorders that result from defects in the development or function of the immune system [47]. Since PIDs can be viewed as developmental defects or malfunctions in immune cells, iPSC-based disease modeling is well suited for their study. Furthermore, because hematopoietic differentiation into white blood cells preferentially derives nonlymphoid cells such as monocytes and neutrophils, cellular defects are easier to model. Among the causative genes for severe congenital neutropenia (SCN), researchers have focused on *elastase, neutrophil expressed (ELANE)* [48]. ELANE-mutated iPSCs showed maturation arrest of neutrophils [49, 50]. Furthermore, Wnt signaling was associated with neutrophil maturation, and mislocalization of neutrophil elastase induced a stress response of the endoplasmic reticulum and cell death [49, 50]. An in vitro iPSC model of SCN with *HCLS1 associated protein X-1 (HAX1)* gene mutation could also recapitulate the neutrophil maturation arrest [51, 52]. Reticular dysgenesis (RD), a severe combined immunodeficiency disease with neutrophil maturation arrest, is caused by mutations in the *adenylate kinase 2 (AK2)* gene [53, 54]. One study established a patient iPSC model of RD, showing that AK2-deficient hematopoietic cells exhibit a reduced intracellular energetic status with increased oxidative stress, which is associated with the impaired neutrophil maturation [55]. The loss of AK2 activity is also associated with the decrease of ATP concentration in the nucleus of hematopoietic progenitor cells [56]. These iPSC-based findings provide unique insights into the relationship between the intracellular bioenergetic status and hematopoietic cell differentiation. In another PID, named Mendelian susceptibility to mycobacterial disease (MSMD), which is caused by *interferon gamma receptor 1 (IFNGR1)* deficiency, iPSC-derived macrophages exhibited previously undescribed defective interferon signaling, providing a novel disease modeling platform [57].

Recently, PSC models of autoimmune diseases have also been established. Autoimmune diseases are a group of diseases caused by dysregulation in acquired immunity. It has been thought that establishing disease models of autoimmune diseases using iPSCs is difficult, because multiple organs and cell types are involved in the pathogenesis and most autoimmune diseases are not associated with a single gene mutation. Sjögren syndrome (SjS) is an autoimmune disease associated with a functional deterioration of glandular cells such as salivary glands and lacrimal glands [58]. In one iPSC model of Sjögren syndrome, dendritic cells could be used to evaluate pathological T cells [59]. iPSCs from RAS-associated autoimmune lymphoproliferative syndrome-like disease (RALD), which is caused by oncogenic *KRAS proto-oncogene, GTPase* (*KRAS*) somatic mutation, have been established [60]. These cells led to the discovery that a specific KRAS pathway is governed by the KRAS mutant. Another group established an iPSC model of Aicardi-Goutières syndrome (AGS), a monogenic autoimmune disease caused by mutations in the *three prime repair exonuclease 1 (TREX1)* gene [61]. Since AGS mainly affects the central nervous system, the study focused on neural lineage cells and showed that the accumulation of certain nucleic acids in and the secretion of neurotoxic cytokines from iPSC-derived astrocytes [62].

In contrast to autoimmune diseases, autoinflammatory syndromes are defined as disorders associated with the dysregulation of innate immunity [63]. Patients with autoinflammatory syndrome suffer from proinflammatory symptoms, such as periodic fever, skin rash, and sterile serositis [64, 65]. iPSC models of autoinflammatory syndrome are feasible, because typical autoinflammatory disorders are monogenic, and the cells responsible for the autoinflammation are usually innate immune cells, which can be robustly differentiated from PSCs. Neonatal-onset multisystem inflammatory disease (NOMID; OMIM #607715, also known as chronic infantile neurological cutaneous and articular syndrome (CINCA)) is a congenital autoinflammatory disease with severe inflammatory symptoms [66, 67]. Approximately 90% of NOMID patients carry mutations in the *NLR family pyrin domain containing 3 (NLRP3)* gene either as a constitutive mutation [68, 69] or somatic mosaicism [70, 71]. Because each iPSC clone is derived from a single somatic cell [72], the establishment of iPSCs from mosaic patients enables the obtainment of both wild-type and mutant clones from an identical genetic background. Indeed, an iPSC model of NOMID in which iPSCs were established from two mosaic patients succeeded to recapitulate distinct proinflammatory phenotypes in mutant iPSC-derived macrophages [73]. In another NOMID-iPSC model, iPSCs were established from a NLRP3-mutation-negative patient [74]. Interestingly, some iPSC clones showed proinflammatory phenotypes while others did not. Whole exome sequencing of the iPSC clones identified a mutation in the *NLR family CARD domain containing 4 (NLRC4)* gene only in diseased clones, indicating the patient had a somatic disease-causing NLRC4 mutation. This case is the first report in which iPSC technology played an essential role in identifying the precise genetic background of a patient.

In another autoinflammatory syndrome named Blau syndrome, which is caused by mutations in the *nucleotide-binding oligomerization domain-containing protein 2 (NOD2)* gene [75, 76], a phenotype analysis of iPSC-derived monocytic cells identified a novel signaling pathway that evokes a proinflammatory response in mutant cells [77].

Some autoinflammatory syndromes are caused by the dysregulation of the proteasome complex and therefore named proteasome-associated autoinflammatory syndrome (PRAAS) [78, 79]. PRAAS patients show an upregulation of the interferon signature with characteristic phenotypes. An iPSC model of Nakajo-Nishimura syndrome (NNS), a member of PRAAS, has been established [18]. Here, mutant-iPSC-derived monocytic cells secreted increased amounts of inflammatory cytokines and chemokines, which was associated with increased oxidative stress. These studies show the validity of iPSC models for dissecting the pathogenesis and understanding the pathophysiology of autoinflammatory syndromes.

2.5 In Vitro Modeling of Gene Therapy Using iPSCs

Research on gene therapies using iPSCs for hematological and immunological disorders have also been conducted. Here, a gene therapy is applied to patient-derived iPSCs, and HSCs induced from the treated iPSCs are transplanted to the patient. The first example of iPSC-based gene therapy was reported by Hanna et al. [80]. In that study, iPSCs were established from mice with sickle cell anemia. The *hemoglobin beta chain complex (Hbb)* gene of the iPSCs was repaired, and the cells were then differentiated into hematopoietic progenitor cells, which were transplanted into diseased mice, which ameliorated the anemia. This study demonstrated the significance of autologous cell transplantation therapy using gene-repaired iPSCs. Furthermore, the recent evolution of genome editing technology using sequence-specific nucleases such as clustered regularly interspaced short palindromic repeats-caspase 9 (CRISPR/Cas9) [81, 82] and transcription activator-like effector nucleases (TALENs) [83, 84] has enabled gene targeting in PSCs that is easier and more efficient. By combining genome editing technology and iPSC technology, now one can generate gene-corrected somatic cells of higher quality.

For iPSC models of hereditary anemia, repair of the β globin gene using sequence-specific nuclease technology has been reported [85–87]. In the case of X-linked severe combined immunodeficiency syndrome (SCID-X1), a PID in which gene therapy has already taken place [88, 89], the *interleukin 2 receptor subunit gamma (IL2RG)* gene in iPSCs established from a X-SCID patient was repaired by genomic editing technology [90]. The in vitro differentiation of repaired iPSCs showed recovery of the differentiation capacity for natural killer cells [90]. Another example is the correction of *pyruvate kinase L/R (PKLR)*, a causative gene of pyruvate kinase deficiency [91, 92]. Genetic correction of the *PKLR* gene using TALENs recovered the energy metabolism of iPSC-derived erythroid progenitor cells, providing a proof of principle of gene therapy for metabolic erythroid disorders [90]. Similar gene therapy models using disease-specific iPSCs derived from patients with chronic granulomatous disease, a typical disorder of phagocytic dysfunction, were also proposed [93–95]. Although most of these studies showed the recovery of pathological phenotypes only in in vitro assays, they clearly showed the proof of concept of future strategy for gene-targeted cell therapy based on iPSC technology.

2.6 Conclusion

Disease-specific iPSCs have allowed scientists to reproduce disease phenotypes in vitro in human cells. Furthermore, iPSCs have helped clarify the molecular mechanisms of various hematopoietic and immunological disorders and advanced relevant drug discovery. Continuous developments in iPSC technology are expected to elaborate and expand current research for more detailed understanding of the pathology, faster drug discovery and repositioning, and more reliable toxicity assays. Overall, the development of novel therapeutic approaches from iPSC-based research will have tremendous benefits for patients with hematopoietic and immunological disorders.

Acknowledgment I thank Dr. Peter Karagiannis (CiRA, Kyoto University) for reading and editing the manuscript.

References

1. Paes B, Moco PD, Pereira CG, Porto GS, de Sousa Russo EM, Reis LCJ, et al. Ten years of iPSC: clinical potential and advances in vitro hematopoietic differentiation. Cell Biol Toxicol. 2017;33(3):233–50.
2. Lim WF, Inoue-Yokoo T, Tan KS, Lai MI, Sugiyama D. Hematopoietic cell differentiation from embryonic and induced pluripotent stem cells. Stem Cell Res Ther. 2013;4(3):71.

3. Slukvin II. Hematopoietic specification from human pluripotent stem cells: current advances and challenges toward de novo generation of hematopoietic stem cells. Blood. 2013;122(25):4035–46.
4. Tan YT, Ye L, Xie F, Beyer AI, Muench MO, Wang J, et al. Respecifying human iPSC-derived blood cells into highly engraftable hematopoietic stem and progenitor cells with a single factor. Proc Natl Acad Sci U S A. 2018;115(9):2180–5.
5. Sugimura R, Jha DK, Han A, Soria-Valles C, da Rocha EL, Lu YF, et al. Haematopoietic stem and progenitor cells from human pluripotent stem cells. Nature. 2017;545(7655):432–8.
6. Suzuki N, Yamazaki S, Yamaguchi T, Okabe M, Masaki H, Takaki S, et al. Generation of engraftable hematopoietic stem cells from induced pluripotent stem cells by way of teratoma formation. Mol Ther. 2013;21(7):1424–31.
7. Niwa A, Heike T, Umeda K, Oshima K, Kato I, Sakai H, et al. A novel serum-free monolayer culture for orderly hematopoietic differentiation of human pluripotent cells via mesodermal progenitors. PLoS One. 2011;6(7):e22261.
8. Yanagimachi MD, Niwa A, Tanaka T, Honda-Ozaki F, Nishimoto S, Murata Y, et al. Robust and highly-efficient differentiation of functional monocytic cells from human pluripotent stem cells under serum- and feeder cell-free conditions. PLoS One. 2013;8(4):e59243.
9. Kaufman DS, Hanson ET, Lewis RL, Auerbach R, Thomson JA. Hematopoietic colony-forming cells derived from human embryonic stem cells. Proc Natl Acad Sci. 2001;98(19):10716–21.
10. Vodyanik MA, Bork JA, Thomson JA, Slukvin II. Human embryonic stem cell–derived $CD34^+$ cells: efficient production in the coculture with OP9 stromal cells and analysis of lymphohematopoietic potential. Blood. 2005;105(2):617–26.
11. Keller G, Kennedy M, Papayannopoulou T, Wiles MV. Hematopoietic commitment during embryonic stem cell differentiation in culture. Mol Cell Biol. 1993;13(1):473–86.
12. Ng ES, Davis RP, Azzola L, Stanley EG, Elefanty AG. Forced aggregation of defined numbers of human embryonic stem cells into embryoid bodies fosters robust, reproducible hematopoietic differentiation. Blood. 2005;106(5):1601–3.
13. Nakamura S, Takayama N, Hirata S, Seo H, Endo H, Ochi K, et al. Expandable megakaryocyte cell lines enable clinically applicable generation of platelets from human induced pluripotent stem cells. Cell Stem Cell. 2014;14(4):535–48.
14. Doulatov S, Vo LT, Chou SS, Kim PG, Arora N, Li H, et al. Induction of multipotential hematopoietic progenitors from human pluripotent stem cells via respecification of lineage-restricted precursors. Cell Stem Cell. 2013;13(4):459–70.
15. Doulatov S, Vo LT, Macari ER, Wahlster L, Kinney MA, Taylor AM, et al. Drug discovery for Diamond-Blackfan anemia using reprogrammed hematopoietic progenitors. Sci Transl Med. 2017;9(376):eaah5645.
16. Haruta M, Tomita Y, Yuno A, Matsumura K, Ikeda T, Takamatsu K, et al. TAP-deficient human iPS cell-derived myeloid cell lines as unlimited cell source for dendritic cell-like antigen-presenting cells. Gene Ther. 2013;20(5):504–13.
17. Haga E, Endo Y, Haruta M, Koba C, Matsumura K, Takamatsu K, et al. Therapy of peritoneally disseminated colon cancer by TAP-deficient embryonic stem cell-derived macrophages in allogeneic recipients. J Immunol. 2014;193(4):2024–33.
18. Honda-Ozaki F, Terashima M, Niwa A, Saiki N, Kawasaki Y, Ito H, et al. Pluripotent stem cell model of Nakajo-Nishimura syndrome untangles proinflammatory pathways mediated by oxidative stress. Stem Cell Reports. 2018;10(6):1835–50.
19. Takata M, Ishiai M, Kitao H. The Fanconi anemia pathway: insights from somatic cell genetics using DT40 cell line. Mutat Res. 2009;668(1–2):92–102.
20. Mehta PA, Tolar J. Fanconi anemia. In: Adam MP, Ardinger HH, Pagon RA, Wallace SE, Bean LJH, Stephens K, et al., editors. GeneReviews((R)). Seattle: University of Washington. GeneReviews is a registered trademark of the University of Washington, Seattle. All rights reserved; 1993.
21. Mamrak NE, Shimamura A, Howlett NG. Recent discoveries in the molecular pathogenesis of the inherited bone marrow failure syndrome Fanconi anemia. Blood Rev. 2017;31(3):93–9.

22. Gari K, Constantinou A. The role of the Fanconi anemia network in the response to DNA replication stress. Crit Rev Biochem Mol Biol. 2009;44(5):292–325.
23. Bagby G. Recent advances in understanding hematopoiesis in Fanconi Anemia. F1000Res. 2018;7:105.
24. Auerbach AD. Fanconi anemia and its diagnosis. Mutat Res. 2009;668(1–2):4–10.
25. Raya A, Rodriguez-Piza I, Guenechea G, Vassena R, Navarro S, Barrero MJ, et al. Disease-corrected haematopoietic progenitors from Fanconi anaemia induced pluripotent stem cells. Nature. 2009;460(7251):53–9.
26. Navarro S, Giorgetti A, Raya A, Tolar J. Induced pluripotency and gene editing in fanconi anemia. Curr Gene Ther. 2017;16(5):321–8.
27. Liu GH, Suzuki K, Li M, Qu J, Montserrat N, Tarantino C, et al. Modelling Fanconi anemia pathogenesis and therapeutics using integration-free patient-derived iPSCs. Nat Commun. 2014;5:4330.
28. Suzuki NM, Niwa A, Yabe M, Hira A, Okada C, Amano N, et al. Pluripotent cell models of fanconi anemia identify the early pathological defect in human hemoangiogenic progenitors. Stem Cells Transl Med. 2015;4(4):333–8.
29. Pastores GM, Hughes DA. Gaucher disease. In: Adam MP, Ardinger HH, Pagon RA, Wallace SE, Bean LJH, Stephens K, et al., editors. GeneReviews((R)). Seattle: University of Washington. GeneReviews is a registered trademark of the University of Washington, Seattle. All rights reserved; 1993.
30. Sgambato JA, Park TS, Miller D, Panicker LM, Sidransky E, Lun Y, et al. Gaucher disease-induced pluripotent stem cells display decreased erythroid potential and aberrant myelopoiesis. Stem Cells Transl Med. 2015;4(8):878–86.
31. Arai S, Miyauchi M, Kurokawa M. Modeling of hematologic malignancies by iPS technology. Exp Hematol. 2015;43(8):654–60.
32. Kumano K, Arai S, Hosoi M, Taoka K, Takayama N, Otsu M, et al. Generation of induced pluripotent stem cells from primary chronic myelogenous leukemia patient samples. Blood. 2012;119(26):6234–42.
33. Bedel A, Pasquet JM, Lippert E, Taillepierre M, Lagarde V, Dabernat S, et al. Variable behavior of iPSCs derived from CML patients for response to TKI and hematopoietic differentiation. PLoS One. 2013;8(8):e71596.
34. Ben-Neriah Y, Daley GQ, Mes-Masson AM, Witte ON, Baltimore D. The chronic myelogenous leukemia-specific P210 protein is the product of the bcr/abl hybrid gene. Science. 1986;233(4760):212–4.
35. Heisterkamp N, Stam K, Groffen J, de Klein A, Grosveld G. Structural organization of the bcr gene and its role in the Ph' translocation. Nature. 1985;315(6022):758–61.
36. Chao MP, Gentles AJ, Chatterjee S, Lan F, Reinisch A, Corces MR, et al. Human AML-iPSCs reacquire leukemic properties after differentiation and model clonal variation of disease. Cell Stem Cell. 2017;20(3):329–44.e7.
37. Gomez Limia CE, Devalle S, Reis M, Sochacki J, Carneiro M, Madeiro da Costa R, et al. Generation and characterization of a human induced pluripotent stem (iPS) cell line derived from an acute myeloid leukemia patient evolving from primary myelofibrosis carrying the CALR 52bp deletion and the ASXL1 p.R693X mutation. Stem Cell Res. 2017;24:16–20.
38. Lee JH, Salci KR, Reid JC, Orlando L, Tanasijevic B, Shapovalova Z, et al. Brief report: human acute myeloid leukemia reprogramming to pluripotency is a rare event and selects for patient hematopoietic cells devoid of leukemic mutations. Stem Cells. 2017;35(9):2095–102.
39. Connelly JP, Kwon EM, Gao Y, Trivedi NS, Elkahloun AG, Horwitz MS, et al. Targeted correction of RUNX1 mutation in FPD patient-specific induced pluripotent stem cells rescues megakaryopoietic defects. Blood. 2014;124(12):1926–30.
40. Boiers C, Richardson SE, Laycock E, Zriwil A, Turati VA, Brown J, et al. A human IPS model implicates embryonic B-myeloid fate restriction as developmental susceptibility to B acute lymphoblastic leukemia-associated ETV6-RUNX1. Dev Cell. 2018;44(3):362–77.e7.

41. Mulero-Navarro S, Sevilla A, Roman AC, Lee DF, D'Souza SL, Pardo S, et al. Myeloid dysregulation in a human induced pluripotent stem cell model of PTPN11-associated juvenile myelomonocytic leukemia. Cell Rep. 2015;13(3):504–15.
42. Bhatnagar N, Nizery L, Tunstall O, Vyas P, Roberts I. Transient abnormal myelopoiesis and AML in down syndrome: an update. Curr Hematol Malig Rep. 2016;11(5):333–41.
43. Yoshida K, Toki T, Okuno Y, Kanezaki R, Shiraishi Y, Sato-Otsubo A, et al. The landscape of somatic mutations in down syndrome-related myeloid disorders. Nat Genet. 2013;45(11):1293–9.
44. Byrska-Bishop M, VanDorn D, Campbell AE, Betensky M, Arca PR, Yao Y, et al. Pluripotent stem cells reveal erythroid-specific activities of the GATA1 N-terminus. J Clin Invest. 2015;125(3):993–1005.
45. Banno K, Omori S, Hirata K, Nawa N, Nakagawa N, Nishimura K, et al. Systematic cellular disease models reveal synergistic interaction of trisomy 21 and GATA1 mutations in hematopoietic abnormalities. Cell Rep. 2016;15(6):1228–41.
46. Kotini AG, Chang CJ, Chow A, Yuan H, Ho TC, Wang T, et al. Stage-specific human induced pluripotent stem cells map the progression of myeloid transformation to transplantable leukemia. Cell Stem Cell. 2017;20(3):315–28.e7.
47. Picard C, Al-Herz W, Bousfiha A, Casanova J-L, Chatila T, Conley ME, et al. Primary immunodeficiency diseases: an update on the classification from the International Union of Immunological Societies Expert Committee for Primary Immunodeficiency 2015. J Clin Immunol. 2015;35(8):696–726.
48. Dale DC, Person RE, Bolyard AA, Aprikyan AG, Bos C, Bonilla MA, et al. Mutations in the gene encoding neutrophil elastase in congenital and cyclic neutropenia. Blood. 2000;96(7):2317–22.
49. Nayak RC, Trump LR, Aronow BJ, Myers K, Mehta P, Kalfa T, et al. Pathogenesis of ELANE-mutant severe neutropenia revealed by induced pluripotent stem cells. J Clin Invest. 2015;125(8):3103–16.
50. Hiramoto T, Ebihara Y, Mizoguchi Y, Nakamura K, Yamaguchi K, Ueno K, et al. Wnt3a stimulates maturation of impaired neutrophils developed from severe congenital neutropenia patient-derived pluripotent stem cells. Proc Natl Acad Sci U S A. 2013;110(8):3023–8.
51. Morishima T, Watanabe K, Niwa A, Hirai H, Saida S, Tanaka T, et al. Genetic correction of HAX1 in induced pluripotent stem cells from a patient with severe congenital neutropenia improves defective granulopoiesis. Haematologica. 2014;99(1):19–27.
52. Pittermann E, Lachmann N, MacLean G, Emmrich S, Ackermann M, Gohring G, et al. Gene correction of HAX1 reversed Kostmann disease phenotype in patient-specific induced pluripotent stem cells. Blood Adv. 2017;1(14):903–14.
53. Pannicke U, Honig M, Hess I, Friesen C, Holzmann K, Rump EM, et al. Reticular dysgenesis (aleukocytosis) is caused by mutations in the gene encoding mitochondrial adenylate kinase 2. Nat Genet. 2009;41(1):101–5.
54. Lagresle-Peyrou C, Six EM, Picard C, Rieux-Laucat F, Michel V, Ditadi A, et al. Human adenylate kinase 2 deficiency causes a profound hematopoietic defect associated with sensorineural deafness. Nat Genet. 2009;41(1):106–11.
55. Rissone A, Weinacht KG, la Marca G, Bishop K, Giocaliere E, Jagadeesh J, et al. Reticular dysgenesis-associated AK2 protects hematopoietic stem and progenitor cell development from oxidative stress. J Exp Med. 2015;212(8):1185–202.
56. Oshima K, Saiki N, Tanaka M, Imamura H, Niwa A, Tanimura A, et al. Human AK2 links intracellular bioenergetic redistribution to the fate of hematopoietic progenitors. Biochem Biophys Res Commun. 2018;497(2):719–25.
57. Neehus A-L, Lam J, Haake K, Merkert S, Schmidt N, Mucci A, et al. Impaired IFNγ-signaling and mycobacterial clearance in IFNγR1-deficient human iPSC-derived macrophages. Stem Cell Reports. 2018;10(1):7–16.
58. Nair JJ, Singh TP. Sjogren's syndrome: review of the aetiology, pathophysiology & potential therapeutic interventions. J Clin Exp Dent. 2017;9(4):e584–e9.

59. Iizuka-Koga M, Asashima H, Ando M, Lai CY, Mochizuki S, Nakanishi M, et al. Functional analysis of dendritic cells generated from T-iPSCs from CD4+ T cell clones of Sjogren's syndrome. Stem Cell Reports. 2017;8(5):1155–63.
60. Kubara K, Yamazaki K, Ishihara Y, Naruto T, Lin H-T, Nishimura K, et al. Status of KRAS in iPSCs impacts upon self-renewal and differentiation propensity. Stem Cell Reports. 2018;11(2):380–94.
61. Crow YJ, Hayward BE, Parmar R, Robins P, Leitch A, Ali M, et al. Mutations in the gene encoding the 3′–5′ DNA exonuclease TREX1 cause Aicardi-Goutieres syndrome at the AGS1 locus. Nat Genet. 2006;38(8):917–20.
62. Thomas CA, Tejwani L, Trujillo CA, Negraes PD, Herai RH, Mesci P, et al. Modeling of TREX1-dependent autoimmune disease using human stem cells highlights L1 accumulation as a source of neuroinflammation. Cell Stem Cell. 2017;21(3):319–31.e8.
63. McDermott MF, Aksentijevich I, Galon J, McDermott EM, Ogunkolade BW, Centola M, et al. Germline mutations in the extracellular domains of the 55 kDa TNF receptor, TNFR1, define a family of dominantly inherited autoinflammatory syndromes. Cell. 1999;97(1):133–44.
64. Masters SL, Simon A, Aksentijevich I, Kastner DL. Horror autoinflammaticus: the molecular pathophysiology of autoinflammatory disease (*). Annu Rev Immunol. 2009;27:621–68.
65. Goldbach-Mansky R. Immunology in clinic review series; focus on autoinflammatory diseases: update on monogenic autoinflammatory diseases: the role of interleukin (IL)-1 and an emerging role for cytokines beyond IL-1. Clin Exp Immunol. 2012;167(3):391–404.
66. Aksentijevich I, Putnam CD, Remmers EF, Mueller JL, Le J, Kolodner RD, et al. The clinical continuum of cryopyrinopathies: novel CIAS1 mutations in North American patients and a new cryopyrin model. Arthritis Rheum. 2007;56(4):1273–85.
67. Prieur AM, Griscelli C, Lampert F, Truckenbrodt H, Guggenheim MA, Lovell DJ, et al. A chronic, infantile, neurological, cutaneous and articular (CINCA) syndrome. A specific entity analysed in 30 patients. Scand J Rheumatol Suppl. 1987;66:57–68.
68. Aksentijevich I, Nowak M, Mallah M, Chae JJ, Watford WT, Hofmann SR, et al. De novo CIAS1 mutations, cytokine activation, and evidence for genetic heterogeneity in patients with neonatal-onset multisystem inflammatory disease (NOMID): a new member of the expanding family of pyrin-associated autoinflammatory diseases. Arthritis Rheum. 2002;46(12):3340–8.
69. Feldmann J, Prieur AM, Quartier P, Berquin P, Certain S, Cortis E, et al. Chronic infantile neurological cutaneous and articular syndrome is caused by mutations in CIAS1, a gene highly expressed in polymorphonuclear cells and chondrocytes. Am J Hum Genet. 2002;71(1):198–203.
70. Tanaka N, Izawa K, Saito MK, Sakuma M, Oshima K, Ohara O, et al. High incidence of NLRP3 somatic mosaicism in patients with chronic infantile neurologic, cutaneous, articular syndrome: results of an International Multicenter Collaborative Study. Arthritis Rheum. 2011;63(11):3625–32.
71. Saito M, Nishikomori R, Kambe N, Fujisawa A, Tanizaki H, Takeichi K, et al. Disease-associated CIAS1 mutations induce monocyte death, revealing low-level mosaicism in mutation-negative cryopyrin-associated periodic syndrome patients. Blood. 2008;111(4):2132–41.
72. Hanna J, Markoulaki S, Schorderet P, Carey BW, Beard C, Wernig M, et al. Direct reprogramming of terminally differentiated mature B lymphocytes to pluripotency. Cell. 2008;133(2):250–64.
73. Tanaka T, Takahashi K, Yamane M, Tomida S, Nakamura S, Oshima K, et al. Induced pluripotent stem cells from CINCA syndrome patients as a model for dissecting somatic mosaicism and drug discovery. Blood. 2012;120(6):1299–308.
74. Kawasaki Y, Oda H, Ito J, Niwa A, Tanaka T, Hijikata A, et al. Pluripotent cell-based phenotypic dissection identifies a high-frequency somatic NLRC4 mutation as a cause of autoinflammation. Arthritis Rheumatol. 2016;69(2):447–59.
75. Blau EB. Familial granulomatous arthritis, iritis, and rash. J Pediatr. 1985;107(5):689–93.
76. Miceli-Richard C, Lesage S, Rybojad M, Prieur AM, Manouvrier-Hanu S, Hafner R, et al. CARD15 mutations in Blau syndrome. Nat Genet. 2001;29(1):19–20.

77. Takada S, Kambe N, Kawasaki Y, Niwa A, Honda-Ozaki F, Kobayashi K, et al. Pluripotent stem cell models of Blau syndrome reveal an IFN-gamma-dependent inflammatory response in macrophages. J Allergy Clin Immunol. 2017;141(1):339–349.e11.
78. McDermott A, Jacks J, Kessler M, Emanuel PD, Gao L. Proteasome-associated autoinflammatory syndromes: advances in pathogeneses, clinical presentations, diagnosis, and management. Int J Dermatol. 2015;54(2):121–9.
79. Brehm A, Liu Y, Sheikh A, Marrero B, Omoyinmi E, Zhou Q, et al. Additive loss-of-function proteasome subunit mutations in CANDLE/PRAAS patients promote type I IFN production. J Clin Invest. 2015;125(11):4196–211.
80. Hanna J, Wernig M, Markoulaki S, Sun CW, Meissner A, Cassady JP, et al. Treatment of sickle cell anemia mouse model with iPS cells generated from autologous skin. Science. 2007;318(5858):1920–3.
81. Cong L, Ran FA, Cox D, Lin S, Barretto R, Habib N, et al. Multiplex genome engineering using CRISPR/Cas systems. Science. 2013;339(6121):819–23.
82. Jinek M, Chylinski K, Fonfara I, Hauer M, Doudna JA, Charpentier E. A programmable dual-RNA-guided DNA endonuclease in adaptive bacterial immunity. Science. 2012;337(6096):816–21.
83. Hockemeyer D, Wang H, Kiani S, Lai CS, Gao Q, Cassady JP, et al. Genetic engineering of human pluripotent cells using TALE nucleases. Nat Biotechnol. 2011;29(8):731–4.
84. Miller JC, Tan S, Qiao G, Barlow KA, Wang J, Xia DF, et al. A TALE nuclease architecture for efficient genome editing. Nat Biotechnol. 2011;29(2):143–8.
85. Park S, Gianotti-Sommer A, Molina-Estevez FJ, Vanuytsel K, Skvir N, Leung A, et al. A comprehensive, ethnically diverse library of sickle cell disease-specific induced pluripotent stem cells. Stem Cell Reports. 2017;8(4):1076–85.
86. Tangprasittipap A, Jittorntrum B, Wongkummool W, Kitiyanant N, Tubsuwan A. Generation of induced pluripotent stem cells from peripheral blood CD34+ hematopoietic progenitors of a 31 year old healthy woman. Stem Cell Res. 2017;20:91–3.
87. Song B, Fan Y, He W, Zhu D, Niu X, Wang D, et al. Improved hematopoietic differentiation efficiency of gene-corrected beta-thalassemia induced pluripotent stem cells by CRISPR/Cas9 system. Stem Cells Dev. 2015;24(9):1053–65.
88. Gaspar HB, Parsley KL, Howe S, King D, Gilmour KC, Sinclair J, et al. Gene therapy of X-linked severe combined immunodeficiency by use of a pseudotyped gammaretroviral vector. Lancet. 2004;364(9452):2181–7.
89. Cavazzana-Calvo M, Hacein-Bey S, de Saint BG, Gross F, Yvon E, Nusbaum P, et al. Gene therapy of human severe combined immunodeficiency (SCID)-X1 disease. Science. 2000;288(5466):669–72.
90. Garate Z, Quintana-Bustamante O, Crane AM, Olivier E, Poirot L, Galetto R, et al. Generation of a high number of healthy erythroid cells from gene-edited pyruvate kinase deficiency patient-specific induced pluripotent stem cells. Stem Cell Reports. 2015;5(6):1053–66.
91. Brunetti P, Puxeddu A, Nenci G, Migliorini E. Congenital non spherocytic haemolytic anaemia due to pyruvate-kinase deficiency. Acta Haematol. 1963;30:88–102.
92. Oski FA, Diamond LK. Erythrocyte pyruvate kinase deficiency resulting in congenital nonspherocytic hemolytic anemia. N Engl J Med. 1963;269:763–70.
93. Dreyer AK, Hoffmann D, Lachmann N, Ackermann M, Steinemann D, Timm B, et al. TALEN-mediated functional correction of X-linked chronic granulomatous disease in patient-derived induced pluripotent stem cells. Biomaterials. 2015;69:191–200.
94. Flynn R, Grundmann A, Renz P, Hanseler W, James WS, Cowley SA, et al. CRISPR-mediated genotypic and phenotypic correction of a chronic granulomatous disease mutation in human iPS cells. Exp Hematol. 2015;43(10):838–48.e3.
95. Laugsch M, Rostovskaya M, Velychko S, Richter C, Zimmer A, Klink B, et al. Functional restoration of gp91phox-oxidase activity by BAC transgenesis and gene targeting in X-linked chronic granulomatous disease iPSCs. Mol Ther. 2016;24(4):812–22.

Chapter 3
iPS Cell Technology for Dissecting Cancer Epigenetics

Hirofumi Shibata and Yasuhiro Yamada

Abstract A dynamic reorganization of epigenetic regulation takes place during cellular reprogramming. Given that reprogramming does not require changes in the underlying genome information, reprogramming technology can be used to actively modify the epigenetic regulation and thus is useful for studying the genome-epigenome relationship. Cancer cells harbor both genetic and epigenetic alterations. Although the causal role of genetic aberrations on cancer development has been well characterized by reverse genetics in vivo, the impact of epigenetic abnormalities remains to be fully understood, especially in vivo. Recent genome-wide sequencing studies have identified frequent mutations at epigenetic modifier genes, thus indicating that epigenetic alterations in cancer could be the consequence of genetic mutations. However, recent studies that utilized reprogramming technology for cancer research have demonstrated cellular context-associated epigenetic regulation that is independent of genetic mutations and plays a critical role on both the development and maintenance of cancer cells. In this review, we propose that reprogramming technology could be a powerful tool for dissecting the role of epigenetic regulation in cancer biology.

Keywords Cancer · Reprogramming technology · Epigenetics · DNA methylation · In vivo reprogramming

H. Shibata
Department of Life Science Frontiers, Center for iPS Cell Research and Application (CiRA), Kyoto University, Kyoto, Japan

Department of Otolaryngology, Gifu University Graduate School of Medicine, Gifu University, Gifu, Japan
e-mail: hsh@gifu-u.ac.jp

Y. Yamada (✉)
Department of Life Science Frontiers, Center for iPS Cell Research and Application (CiRA), Kyoto University, Kyoto, Japan

Division of Stem Cell Pathology, Center for Experimental Medicine and Systems Biology, Institute of Medical Science, University of Tokyo, Tokyo, Japan
e-mail: yasu@ims.u-tokyo.ac.jp

H. Inoue, Y. Nakamura (eds.), *Medical Applications of iPS Cells*, Current Human Cell Research and Applications,
https://doi.org/10.1007/978-981-13-3672-0_3

Abbreviations

AML	Acute myeloid leukemia
DMRs	Differentially methylated regions
ESCs	Embryonic stem cells
GBM	Glioblastoma
iPCCs	Induced pluripotent cancer cells
iPSCs	Induced pluripotent stem cells
LOH	Loss of heterogeneity
NT-ESCs	Nuclear transferred embryonic stem cells
PDAC	Pancreatic ductal adenocarcinoma
PRC2	Polycomb repressive complex 2
ROS	Reactive oxygen species
RTCs	Reprogrammed tumor cells
SAM	*S*-adenosylmethyonine
SCNT	Somatic cell nuclear transfer
TPCs	Tumor-propagating cells

3.1 Epigenetic Alterations in Cancer

It is known that an accumulation of genetic mutations causes the development of cancer [1, 2]. Indeed, genome-wide sequencing analyses have unveiled a number of oncogenic mutations in various types of cancer. Studies using reverse genetics have revealed the functional significance of oncogenic mutations by introducing these mutations in rodents and reporting the development of cancer types seen in human patients. However, in addition to genetic aberrations, many studies have revealed that cancer development is accompanied by epigenetic modifications that differ from those in normal cells [3–5].

Various types of cancers harbor epigenetic alterations including aberrant DNA methylation and altered histone modifications. Epigenetic changes in human cancer were first described as a reduction of global DNA methylation levels [6]. Particularly, repeat sequences that are heavily methylated in normal cells are often hypomethylated in cancer cells. Mechanistically, such hypomethylation has been connected with genomic instability. Embryonic stem cells (ESCs) deficient for *Dnmt1*, a maintenance DNA methyltransferase, exhibit genome-wide hypomethylation and eventually show chromosomal instability and elevated mutation rates [7]. In addition, forced reduction of genome-wide DNA methylation using *Dnmt1* hypomorphic alleles in mice promotes chromosomal instability and accelerates T-cell lymphoma development as well as sarcoma formation in *Nf1/p53* compound mutant background [8, 9]. Collectively, these studies suggest that a genome-wide reduction in DNA methylation promotes cancer development through genomic instability.

In contrast to the above cancer-promoting effects, reduced DNA methylation has been reported to suppress cancer development in other contexts. A forced reduction of DNA methylation by the heterozygous *Dnmt1* mutation in conjunction with 5-aza-2′-deoxycytidine treatment efficiently suppressed the development of intestinal tumors in *Apc*$^{min/+}$ mouse model [10]. Additionally, opposing effects of reduced DNA methylation have been demonstrated in the same *Apc*$^{min/+}$ mouse. Namely, a reduction of DNA methylation in *Apc*$^{min/+}$ mice suppressed macroscopic colon tumors but promoted the development of early microscopic lesions (microadenomas) in the colon and induced liver tumors [11]. Microadenomas in DNA-hypomethylated *Apc*$^{min/+}$ mice often showed a loss of heterogeneity (LOH) in *Apc* locus, which supports the notion that the chromosomal instability caused by the reduced DNA methylation promoted microadenoma formation. In contrast, a recent study showed that DNA hypomethylation-mediated suppression of colon tumorigenesis is associated with the induction of differentiation in colon tumor cells through the upregulation of *Cdx1* and *Cdx2* genes, which are key transcription factors in intestinal differentiation [12]. All together, these studies suggest that altered DNA methylation works differently depending on the cellular context and stage of tumorigenesis.

It has also been reported that conditional deletion of *Dnmt3b*, a de novo DNA methyltransferase, inhibits the development of intestinal tumors, while the forced expression of *Dnmt3b* promotes the development of colon tumors [13, 14], indicating that de novo DNA methylation plays a role in the promotion stage of colon tumorigenesis. However, the precise mechanisms for how de novo methylation promotes tumorigenesis remain unclear.

It is expected that aberrant de novo methylation results in DNA hypermethylation. Notably, DNA hypermethylation has been observed in the promoter region of CpG islands of tumor suppressor genes [15], such as *APC*, *p16 ink4A*, *BRCA1*, and *RB*, and associated with the promotion of cancer development [16–18]. The current consensus holds that DNA hypermethylation promotes cancer development through the aberrant silencing of various tumor suppressor genes in diverse cancer types.

Abnormal epigenetic modifications are similarly observed at differentially methylated regions (DMRs) of imprinted genes, which are expressed in a parent-of-origin-dependent manner. Among them, aberrant DNA hypermethylation at *IGF2-H19* locus is often observed in various types of childhood cancers such as Wilms' tumor, rhabdoid tumor, and hepatoblastoma [19–21]. *H19* DMR hypermethylation results in the biallelic expression of *IGF2*, resulting in an increased dosage. In normal tissue, the proper expression of *IGF2* is important for balance between the proliferation and the differentiation of tissue stem/progenitor cells, thus suggesting that increased *IGF2* seems to be tightly connected with cancer initiation and progression. Indeed, patients of Beckwith-Wiedemann syndrome with biallelic expression of *IGF2* gene often exhibit a substantially increased risk of cancer including Wilms' tumor development, indicating the causal relationship between epigenetic dysregulation and augmented risk of cancer [22].

3.2 Cause of Epigenetic Alterations in Cancer

3.2.1 Gene Mutations at Epigenetics-Related Genes

Recent genome-wide sequencing studies have elucidated that epigenetic modifier genes are often mutated in various types of cancers [23, 24]. For example, genes which encode components of SWI/SNF chromatin remodeling complexes are often mutated in cancers. Among them, the deletion of *SMARCB1* is related to pediatric rhabdoid tumor [25, 26], which exhibits defects in cellular differentiation, suggesting that aberrant chromatin remodeling causes cancer development, presumably through impaired differentiation. Another member of SWI/SNF chromatin remodeling complexes, AT-rich interaction domain 1A (*ARID1A*), is mutated in over half of ovarian clear cell carcinomas [27, 28]. In addition, *DNMT3A*, a de novo methyltransferase gene, is frequently mutated in acute myeloid leukemia (AML) and T-cell lymphoma [23, 29]. Notably, patients with *DNMT3A* mutation have poor prognosis in AML and T-cell lymphoma [30], thus indicating that *DNMT3A* mutation has prognostic value. On the other hand, *TET2*, which is associated with 5hmC production and DNA demethylation, is recurrently mutated in myeloid malignancies and other hematological disorders [31]. Both activating and inactivating mutations of *EZH2*, which encodes a catalytic subunit of polycomb repressive complex 2 (PRC2), have been identified in certain types of cancers [32–34]. Collectively, emerging evidence in cancer genome sequencing analyses has revealed the substantial impact of genetic mutations on altered epigenetic modifications across diverse cancer types (Fig. 3.1).

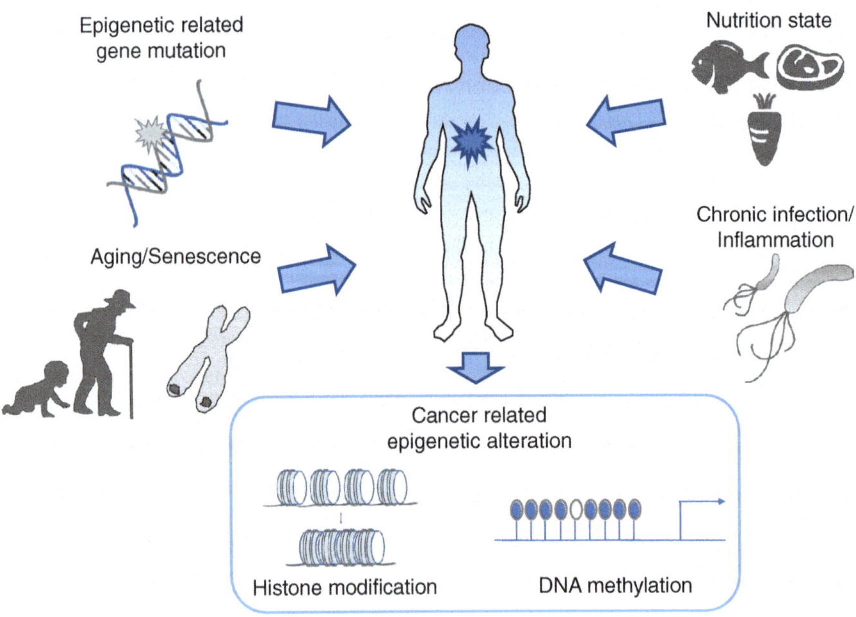

Fig. 3.1 Cause of epigenetic alterations in cancer

3.2.2 Environmental Factor-Mediated Modulation of Epigenetic Landscape

In addition to genetic aberrations, environmental factors also affect epigenetic modifications. Nutrition is one example that influences epigenetic modifications. S-adenosylmethionine (SAM), which works as a methyl donor when DNA and histone are replicated, is synthesized through a carbon metabolism pathway that consists of folic acid, methionine, and vit B12. Defects in the supply of folic acid are thought to affect the availability of methyl groups for DNA methylation. An experimental mouse study indicated that folate depletion during pregnancy and lactation period causes persistent DNA hypomethylation in the small intestine even in adults, and that this effect is not modulated by folate supply after weaning, indicating the memory of epigenetic abnormalities [35]. In addition, adequate folic acid supplementation during maternal but not postweaning significantly reduced the incidence of colorectal cancer, which is accompanied by increased global DNA methylation levels [36]. These data suggest that nutrition affects epigenetic modifications that are associated with cancer development (Fig. 3.1).

Recent studies have also suggested that aging and cellular senescence could influence epigenetic modifications (Fig. 3.1). Early studies observed a progressive depletion of 5mC in senescent fibroblasts and aged tissues [37, 38]. Intriguingly, in a study of human monozygotic twins, significant DNA methylation discordance was observed at multiple loci in older twin pairs [39], which indicates that aging in different environments can cause distinct epigenetic modifications even under the identical genetic background. In relation to cancer, intestinal mucosa acquires remarkable hypermethylation in the promoter region of *IGF2* imprint gene during aging and carcinogenesis, which provides a potential link between aberrant methylation and aging-associated disease [40, 41]. Together, genome-wide studies in aged cells have elucidated a stochastic modification of DNA methylation that reflects an imperfect maintenance of epigenetic marks. Therefore, it is possible that epigenetic mosaicism in aged stem cells may potentially decrease tissue plasticity and induce focal proliferative lesions that have a higher risk for cancer development [42].

Chronic infections and inflammations too have been indicated to promote cancer development. *Helicobacter pylori* infection, which is closely associated with gastric cancer development in both humans and rodents, causes abnormal DNA methylation in the CpG islands of various genes, including *p16 ink 4A* tumor suppressor gene [43] (Fig. 3.1). In addition, chronic inflammation in patients with ulcerative colitis or Barrett's esophagus, who have an increased risk for cancer development, causes higher methylation levels at the CpG island of *p16 ink 4A* and *RUNX3* genes [43, 44].

It has been shown that *reactive oxygen species* (*ROS*) are key signaling molecules in *inflammation-associated diseases*. Notably, ROS and cytokines induce epigenetic abnormalities. ROS-mediated damage of DNA/chromatin recruits the silencing protein complex, which is composed of *DNMT1*, *DNMT3*, and polycomb-related proteins. The complex induces epigenetic silencing including aberrant methylation at the CpG island and affects gene expression profiles [45]. Notably, these epigenetic alterations are similarly observed in cancer cells, suggesting that environmental factor-mediated epigenetic alterations may contribute to cancer development [46]. Considering that these environmental factors are independent of genetic abnormalities in cancer cells, such epigenetic alterations could be bona fide abnormalities that promote cancer development. However, given that genotype is tightly coupled with epigenotype, direct evidence for the role of environmental factor-mediated epigenetic alterations on cancer development is limited, especially in vivo.

3.3 Relationship Between Reprogramming and Cancer Development

The landmark study by Takahashi and Yamanaka revealed that certain transcription factors that are highly expressed in ESCs confer the gene expression profiles and the epigenetic landscape of pluripotent stem cells on differentiated cells. The transient induction of reprogramming transcription factors (*Oct3/4*, *Sox2*, *Klf4*, and *cMyc*) was able to convert somatic cells to ESC-like induced pluripotent stem cells (iPSCs) [47, 48]. It is interesting to note that the reprogramming process toward the pluripotent state shares a number of properties with oncogenic transformations, including the acquisition of self-renewing potential and the loss of original somatic cell identity. In addition, the acquisition of stem cell properties is often observed during carcinogenesis in many organs. The initial event of somatic cell reprogramming includes a loss of the cellular identity that is stably maintained through an epigenetic mechanism in the original somatic cells [49, 50]. Similarly, oncogenic transformation is often accompanied by a loss of cellular identity of the cell of origin or by dedifferentiation. Notably, the degree of dedifferentiation is often correlated with a worse outcome of the disease [51].

It has been shown that some cancers express higher levels of reprogramming factors and that each reprogramming factor has functions related to cancer development. c-*Myc* is known to be a major oncogene in various types of cancer. The tumorigenic role of *Oct3/4* was described in germ cell tumors [52]. In addition, the ectopic expression of *Oct3/4* resulted in the expansion of progenitor cells and led to a cancer-like phenotype in mouse epithelial tissue in vivo [53]. Overexpression of *Nanog* promotes breast cancer tumorigenesis in conjunction with *Wnt* signaling in mouse mammary glands [54]. Similarly, *Sox2* is frequently overexpressed in various types of cancers including esophagus cancer. Moreover, *Sox2* controls the tumor initiation and stemlike function of squamous cell carcinomas [55]. Taken together, these results suggest that overexpressed reprogramming factors may initiate and/or promote cancer through the reprogramming of somatic cells into a progenitor-like state.

3.4 Application of Reprogramming Technologies for Cancer Research

3.4.1 *Cancer Cell Reprogramming*

As mentioned above, many cancers harbor gene mutations in epigenetic modifier genes. In addition, given that transcription factor-mediated gene regulations are tightly connected with epigenetic modifications, an altered transcriptional network caused by genetic mutations may affect epigenetic modifications in cancer cells. As a result, the majority of epigenetic alterations in cancers could be the consequence of dysfunctional epigenetic modifiers and/or an altered transcriptional profile that are caused by genetic abnormalities. It remains unclear whether bona fide epigenetic abnormalities have a causal impact on cancer development.

Reprogramming technologies can induce the reorganization of gene regulatory systems and cause alterations in somatic cell fate. Therefore, these technologies could be used to actively alter epigenetic regulations while preserving genomic information. Indeed, taking advantage of reprogramming technology, previous studies have revealed the importance of epigenetic regulation in cancer development, as described below.

The first studies that reprogrammed a mammalian cancer nucleus did so by somatic cell nuclear transfer (SCNT). Notably, oocytes that received cancer nuclei were subsequently able to form a blastocyst-like structure [56, 57]. These studies suggested that some cancer nuclei are permissive for the initial development of cancer, indicating plasticity of the cancer cell genome. Moreover, it was possible to establish NT-ESCs from cloned blastocyst using *HRAS*-induced melanoma nuclei [57]. Consist with the notion that cellular reprogramming preserves genomic information of the original cells, melanoma NT-ESCs harbor genetic abnormalities common with the parental melanoma cells. Although melanoma NT-ESCs were able to contribute to adult chimeric mice, these mice developed secondary melanoma with a shorter latency and a higher penetrance than the original melanoma model (Fig. 3.2). Secondary melanoma development even after reprogramming into the pluripotent stem cell state highlights the significance of genetic abnormalities in cancer development.

3.4.2 *Cell-Type-Specific Carcinogenesis*

Each cancer type is often associated with mutations at a specific set of genes. For example, *KRAS* mutation is frequently detectable in pancreatic ductal adenocarcinoma, and *HER2* amplification is preferentially observed in breast cancers. In addition, familiar adenomatous polyposis patients show a mutation at the *APC* gene and develop cancers predominantly in the colon, even though *APC* mutation is present in cells throughout the patient's whole body [58]. These observations suggest that

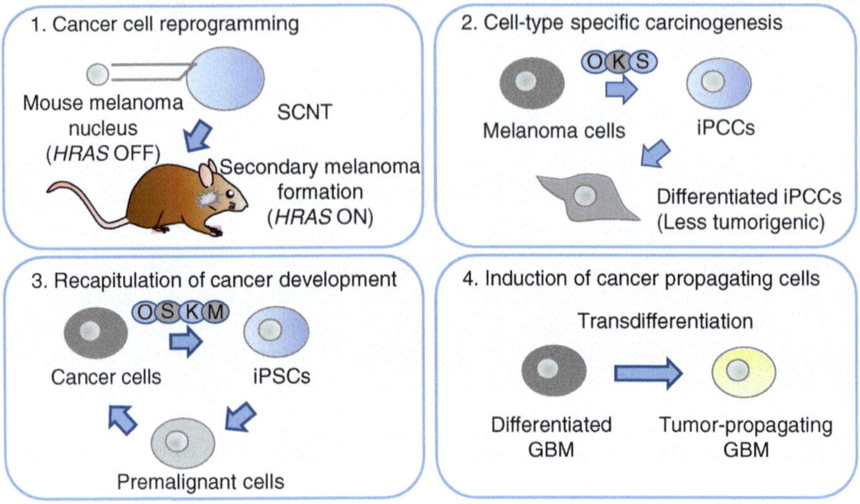

Fig. 3.2 Application of reprogramming technologies to cancer research. (*1*) SCNT has been used to reprogram a cancer nucleus in order to establish NT-ESCs. Note that melanoma NT-ESC chimeric mice developed secondary melanomas with shorter latency, which highlights the importance of genetic abnormalities. (*2*) *BRAF*-mutated mouse melanoma cells were reprogrammed into induced iPCCs. iPCCs showed impaired tumorigenic potential when they were differentiated into mesodermal lineage, which suggests the importance of the cellular context, which is mainly governed by epigenetic regulation, on cancer development. (*3*) Human cancer cell-derived iPSCs may be a unique platform to evaluate the process of carcinogenesis in vivo. Pancreatic cancer progression was recapitulated in teratomas. (*4*) Direct reprogramming by the transient expression of transcription factors induced tumor-propagating cancer cells from differentiated cancer cells. This finding suggests that heterogeneity within a tumor mass and the cancer propagation potential could be governed by epigenetic regulation

the genetic mutations require a specific cellular context in order to exhibit cancer-initiating properties and that cell-type-associated epigenetic regulation may influence tumorigenesis. Similarly, the differentiation status determines responses to *MYC* expression in osteosarcoma cells, suggesting that cellular differentiation-associated epigenetic regulation affects cancer development. Consistent with this idea, differentiation-inducing therapy successfully causes the regression of some tumors, such as all-trans retinoic acid treatment in acute promyelocytic leukemia patients [59]. Given that somatic cells are reprogrammable into iPSCs that are capable of redifferentiating into multiple cell types through global reorganization of the epigenetic regulation, cancer cell reprogramming technology is applicable to study the relationship between cell-type-associated epigenetic regulation and the cancer genome.

Indeed, CML-derived human iPSCs are insensitive to tyrosine kinase inhibitors that target *BCR-ABL* despite the cells expressing *BCR-ABL* gene. Intriguingly, drug sensitivity was found when CML-iPSCs were differentiated into hematopoietic cells [60]. Given that iPSCs and hematopoietic lineage cells share the same genetic context, these results strongly indicate that drug sensitivity in CML depends on the cell-differentiation status governed by epigenetic regulation. Similarly, an experimental study using induced pluripotent cancer cells (iPCCs) from *BRAF* ^{V600E}-mutated melanomas showed that iPCC-derived progeny displays an increased resistance against MAPK-targeted therapies despite these cells harboring the same *BRAF* mutation and signaling activities as paternal melanoma cells [61]. The study also showed iPCCs acquired non-tumorigenic traits after differentiation, suggesting that the tumorigenic potential is affected by the cellular context, which again is mainly governed by epigenetic regulation (Fig. 3.2).

A recent study using *Apc*$^{\text{min}/+}$ mouse model containing a reprogramming system [62] also provides in vivo evidence for the critical role of the cellular context on the cancer genome. In that study, reprogrammed tumor cells (RTCs) were established from macroscopic colon tumor cells in *Apc*$^{\text{min}/+}$ mice by inducing reprogramming factors. Notably, the majority of genes that were affected by *Apc* mutation in RTCs did not overlap with genes affected by *Apc* mutation in differentiated intestinal epithelial cells. This result further supports the notion that cell-type-associated epigenetic regulation plays a critical role on the effects of genetic mutations.

3.4.3 Recapitulation of Human Cancer Progression In Vivo

Mouse models harboring specific oncogenic mutations have been used to investigate the effects of genetic mutations on cancer development. However, biological differences between humans and rodents may interfere with complete understanding of cancer development in humans. iPSCs derived from human cancer cells offer an alternative model for human cancer development. One study successfully established iPSC-like cells from human pancreatic ductal adenocarcinoma (PDAC) harboring *KRAS* mutation, which can be maintained under the low-level expression of exogenous reprogramming factors. Importantly, PDAC-derived iPSC-like cells were capable of differentiating into the three germ layers, suggesting that they have pluripotency [63] (Fig. 3.2). Interestingly, in teratomas from PDAC-derived iPSCs, the sequential progression of PDAC through pancreatic intraepithelial neoplasia, a precancerous lesion, was observed. These results suggest that cancer-derived iPSCs may be a unique platform for investigating human cancer progression in vivo, although further investigation is needed to expand this approach to other cancer types.

3.4.4 Hierarchy of Heterogeneous Cancer Cells

During mammalian development, cell fate decisions are dictated by coordinated regulation of tissue-specific master transcription factors and underlying epigenetic modifications. In addition, it is well established that epigenetic regulation during cell fate determination can be overridden by the artificial induction of core transcription factor cocktails, which was demonstrated by the derivation of iPSCs or by the induction of direct lineage conversion [47, 48]. Somatic stem cells have the ability to initiate the tissue-specific hierarchical differentiation that causes tissue heterogeneity. Certain malignant tumors also have cellular hierarchies and contain tumor-propagating cells (TPCs). In glioblastomas, a subset of stemlike TPCs appears to drive tumor progression and is suggested as causative for therapeutic resistance. Notably, a core set of neurodevelopmental transcription factors (*POU3F2*, *SOX2*, *SALL2*, and *OLIG2*) is essential for glioblastoma (GBM) propagation [64] (Fig. 3.2). Moreover, forced but transient expression of the transcription factor cocktail converted differentiated GBM cells into TPCs. This study indicated that intra-tumoral heterogeneity and TPC properties are regulated by an epigenetic landscape and also suggested that TPC properties are not determined by genetic information in glioblastomas. Similarly, human melanoma cells revealed profound transcriptional variability at the single-cell level that causes resistance to chemotherapy [65]. Notably, there were no genetic differences among observed melanoma cells, but there was transcriptional variability, indicating the involvement of epigenetic dysregulation in the heterogeneity of cancer cells.

3.5 Cancer Research Using In Vivo Reprogramming Technology

Considering the number of shared aspects of the reprogramming process and cancer development, which include the acquisition of self-renewal capacity and impaired terminal differentiation, reprogramming technology in vivo may provide a unique experimental platform to elucidate the impact of epigenetic regulation on cancer development. The implication of in vivo reprogramming was first shown as teratoma formation in mice containing lentivirus-mediated transgenic alleles for doxycycline (Dox)-inducible reprogramming factors [66, 67]. These transgenic mice often developed teratomas containing differentiated cells of the three germ layers, presumably because of leaky expressions of the reprogramming factors. These observations suggested that differentiated somatic cells can be reprogrammed into the pluripotent state in vivo.

Ohnishi et al. generated in vivo reprogrammable mice in which reprogramming alleles are targeted into fixed loci and the expression of reprogramming factors can be visualized by mCherry fluorescence [68] (Fig. 3.3). In this system, a 28-day induction of reprogramming factors resulted in teratomas that contained iPSCs in various organs, indicating successful reprogramming in vivo. In contrast, a shorter transient induction of the reprogramming factors often caused reversible dysplastic lesions in the pancreas, liver, and intestine, reflecting the existence of epigenetic memory in somatic cells. Surprisingly, these mice showed irreversible cancer development in multiple organs after a prolonged (1 week) but transient expression of reprogramming factors (Fig. 3.3). Despite the absence of mCherry expression, the cancer cells exhibited self-renewing and invasion capacity. Notably, cancers in the kidney resembled Wilms' tumor, which is one of the most common kidney cancers of children. These kidney cancers showed traits that were intermediate between

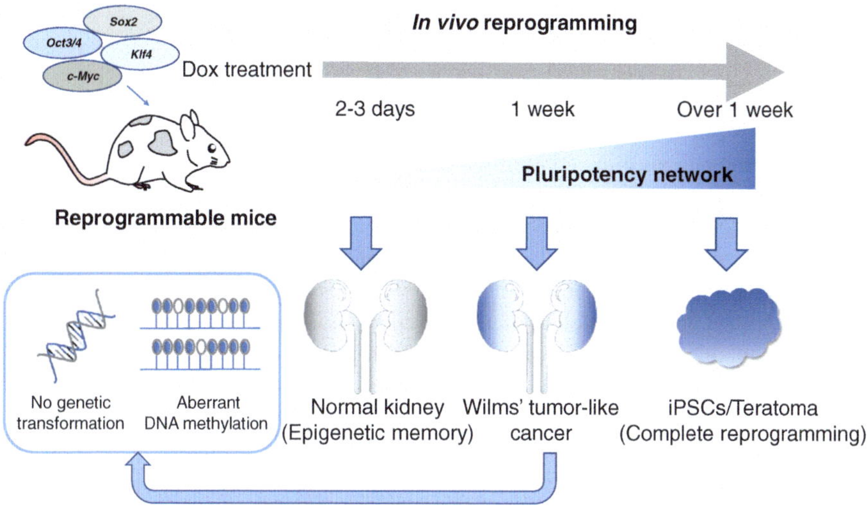

Fig. 3.3 Cancer research using in vivo reprogramming technology

differentiated normal kidney cells and pluripotent stem cells, suggesting a partial reprogramming state. Consistently, these cancers exhibited an activation of ESC-CORE and ESC-MYC modules, which indicates partial acquisition of the pluripotency-related transcription network. Although *WT1* and *WTX* genes are mutated in a subset of human Wilms' tumors, recent whole-genome sequencing analyses revealed that several cases show no recurrent mutations at oncogene-related genes [69]. Importantly, the absence of recurrent mutations was reported in other childhood cancers. Given that cellular reprogramming does not require any particular genetic alterations, it is possible that the development of some childhood cancers may mainly depend on epigenetic dysfunction rather than genetic abnormalities.

3.6 Technical Problems in the Application of Reprogramming Technology to Cancer Research

It is widely accepted that cancer cells are refractory to cellular reprogramming. Particularly, it is known that the establishment of iPSCs from cancer cell lines is extremely difficult, suggesting that strong selection may occur during the reprogramming of cancer cells. In addition, cancer-related genetic mutations may have positive or negative effects on cellular reprogramming. For example, *TP53* inactivation is reported to have a positive effect, whereas silencing of *Ink4a/Arf* locus has a rate-limiting effect on iPSC derivation [70, 71]. Considering that cellular reprogramming is a cloning process, these facts raise the possibility that iPSCs derived from cancer cells may have a selected genetic landscape during the iPSC derivation. It is necessary to carefully consider such possibilities when interpreting the results obtained with cancer-derived iPSCs.

3.7 Future Perspective

Accumulating evidence suggests that epigenetic dysfunction has a causal impact on cancer development. Although the majority of cancers display genetic abnormalities, cancer reprogramming technology has revealed cell-type specificity of cancer development and highlighted the importance of coupling between cellular context and epigenetic regulation. Moreover, this technology has provided in vivo proof of concept for epigenetics-driven cancer development. It is noteworthy that reprogramming is a stepwise process. In the initial phase of reprogramming, a loss of cellular identity occurs, which is characterized by the repression of somatic enhancers. Subsequently, gradual activation of the pluripotency signature occurs without affecting genomic information [49, 50]. Notably, these processes are similarly observed in a subset of cancers, which raises the possibility that common machinery is involved in both reprogramming and the multistep progression of cancer. Collectively, in vivo reprogramming technology holds promise to elucidate the role of bona fide abnormalities in epigenetic regulation on the stepwise development of cancer. Future investigations with this novel epigenetics-editing technology will bring us an innovative perspective on the importance of epigenetic regulation in cancer development in vivo. Recently, more advanced technologies were developed to alter locus-specific epigenetic modifications [72, 73]. Targeting bona fide epigenetic abnormalities with various epigenetics-editing technologies may provide a novel and efficient strategy to treat cancer patients.

Mutations of epigenetic modifier genes are often observed in various types of cancers. Nutrition status, aging/senescence, and chronic infection/inflammation also affect epigenetic regulation. Epigenetic changes caused by these factors have been implicated in cancer development.

In vivo reprogramming technology revealed the importance of epigenetic regulation on cancer development. Chimeric mice in which reprogramming factors can be induced developed iPSCs/teratoma after Dox treatment for 1 month. In contrast, premature termination of in vivo reprogramming caused the development of kidney cancers that resemble Wilms' tumor. These observations suggest that reprogramming-related epigenetic dysregulation may induce some types of cancer development.

References

1. Fearon ER, Hamilton SR, Vogelstein B. Clonal analysis of human colorectal tumors. Science (New York, NY). 1987;238:193–7.
2. Nowell PC. The clonal evolution of tumor cell populations. Science (New York, NY). 1976;194:23–8.
3. Jones PA, Baylin SB. The fundamental role of epigenetic events in cancer. Nat Rev Genet. 2002;3:415–28. https://doi.org/10.1038/nrg816.
4. Hattori N, Ushijima T. Compendium of aberrant DNA methylation and histone modifications in cancer. Biochem Biophys Res Commun. 2014;455:3–9. https://doi.org/10.1016/j.bbrc.2014.08.140.

5. Kondo Y, Katsushima K, Ohka F, Natsume A, Shinjo K. Epigenetic dysregulation in glioma. Cancer Sci. 2014;105:363–9.
6. Feinberg AP, Vogelstein B. Hypomethylation distinguishes genes of some human cancers from their normal counterparts. Nature. 1983;301:89–92.
7. Chen RZ, Pettersson U, Beard C, Jackson-Grusby L, Jaenisch R. DNA hypomethylation leads to elevated mutation rates. Nature. 1998;395:89–93. https://doi.org/10.1038/25779.
8. Gaudet F, et al. Induction of tumors in mice by genomic hypomethylation. Science (New York, NY). 2003;300:489–92. https://doi.org/10.1126/science.1083558.
9. Eden A, Gaudet F, Waghmare A, Jaenisch R. Chromosomal instability and tumors promoted by DNA hypomethylation. Science (New York, NY). 2003;300:455. https://doi.org/10.1126/science.1083557.
10. Laird PW, et al. Suppression of intestinal neoplasia by DNA hypomethylation. Cell. 1995;81:197–205.
11. Yamada Y, et al. Opposing effects of DNA hypomethylation on intestinal and liver carcinogenesis. Proc Natl Acad Sci U S A. 2005;102:13580–5. https://doi.org/10.1073/pnas.0506612102.
12. Hatano Y, et al. Reducing DNA methylation suppresses colon carcinogenesis by inducing tumor cell differentiation. Carcinogenesis. 2015;36:719–29. https://doi.org/10.1093/carcin/bgv060.
13. Lin H, et al. Suppression of intestinal neoplasia by deletion of Dnmt3b. Mol Cell Biol. 2006;26:2976–83. https://doi.org/10.1128/mcb.26.8.2976-2983.2006.
14. Linhart HG, et al. Dnmt3b promotes tumorigenesis in vivo by gene-specific de novo methylation and transcriptional silencing. Genes Dev. 2007;21:3110–22. https://doi.org/10.1101/gad.1594007.
15. Herman JG, Baylin SB. Gene silencing in cancer in association with promoter hypermethylation. N Engl J Med. 2003;349:2042–54. https://doi.org/10.1056/NEJMra023075.
16. Belinsky SA, et al. Aberrant methylation of p16(INK4a) is an early event in lung cancer and a potential biomarker for early diagnosis. Proc Natl Acad Sci U S A. 1998;95:11891–6.
17. Das PM, Singal R. DNA methylation and cancer. J Clin Oncol. 2004;22:4632–42. https://doi.org/10.1200/jco.2004.07.151.
18. Esteller M, et al. Promoter hypermethylation and BRCA1 inactivation in sporadic breast and ovarian tumors. J Natl Cancer Inst. 2000;92:564–9.
19. Rainier S, et al. Relaxation of imprinted genes in human cancer. Nature. 1993;362:747–9. https://doi.org/10.1038/362747a0.
20. Ogawa O, et al. Relaxation of insulin-like growth factor II gene imprinting implicated in Wilms' tumour. Nature. 1993;362:749–51. https://doi.org/10.1038/362749a0.
21. Bjornsson HT, et al. Epigenetic specificity of loss of imprinting of the IGF2 gene in Wilms tumors. J Natl Cancer Inst. 2007;99:1270–3. https://doi.org/10.1093/jnci/djm069.
22. DeBaun MR, et al. Epigenetic alterations of H19 and LIT1 distinguish patients with Beckwith-Wiedemann syndrome with cancer and birth defects. Am J Hum Genet. 2002;70:604–11. https://doi.org/10.1086/338934.
23. Ley TJ, et al. DNMT3A mutations in acute myeloid leukemia. N Engl J Med. 2010;363:2424–33. https://doi.org/10.1056/NEJMoa1005143.
24. Schwartzentruber J, et al. Driver mutations in histone H3.3 and chromatin remodelling genes in paediatric glioblastoma. Nature. 2012;482:226–31. https://doi.org/10.1038/nature10833.
25. Versteege I, et al. Truncating mutations of hSNF5/INI1 in aggressive paediatric cancer. Nature. 1998;394:203–6. https://doi.org/10.1038/28212.
26. Biegel JA, et al. Germ-line and acquired mutations of INI1 in atypical teratoid and rhabdoid tumors. Cancer Res. 1999;59:74–9.
27. Jones S, et al. Frequent mutations of chromatin remodeling gene ARID1A in ovarian clear cell carcinoma. Science (New York, NY). 2010;330:228–31. https://doi.org/10.1126/science.1196333.
28. Wiegand KC, et al. ARID1A mutations in endometriosis-associated ovarian carcinomas. N Engl J Med. 2010;363:1532–43. https://doi.org/10.1056/NEJMoa1008433.

29. Couronne L, Bastard C, Bernard OA. TET2 and DNMT3A mutations in human T-cell lymphoma. N Engl J Med. 2012;366:95–6. https://doi.org/10.1056/NEJMc1111708.
30. Ribeiro AF, et al. Mutant DNMT3A: a marker of poor prognosis in acute myeloid leukemia. Blood. 2012;119:5824–31. https://doi.org/10.1182/blood-2011-07-367961.
31. Delhommeau F, et al. Mutation in TET2 in myeloid cancers. N Engl J Med. 2009;360:2289–301. https://doi.org/10.1056/NEJMoa0810069.
32. McCabe MT, et al. EZH2 inhibition as a therapeutic strategy for lymphoma with EZH2-activating mutations. Nature. 2012;492:108–12. https://doi.org/10.1038/nature11606.
33. Morin RD, et al. Somatic mutations altering EZH2 (Tyr641) in follicular and diffuse large B-cell lymphomas of germinal-center origin. Nat Genet. 2010;42:181–5. https://doi.org/10.1038/ng.518.
34. Ernst T, et al. Inactivating mutations of the histone methyltransferase gene EZH2 in myeloid disorders. Nat Genet. 2010;42:722–6. https://doi.org/10.1038/ng.621.
35. McKay JA, Waltham KJ, Williams EA, Mathers JC. Folate depletion during pregnancy and lactation reduces genomic DNA methylation in murine adult offspring. Genes Nutr. 2011;6:189–96. https://doi.org/10.1007/s12263-010-0199-1.
36. Sie KK, et al. Effect of maternal and postweaning folic acid supplementation on colorectal cancer risk in the offspring. Gut. 2011;60:1687–94. https://doi.org/10.1136/gut.2011.238782.
37. Wilson VL, Jones PA. DNA methylation decreases in aging but not in immortal cells. Science (New York, NY). 1983;220:1055–7.
38. Wilson VL, Smith RA, Ma S, Cutler RG. Genomic 5-methyldeoxycytidine decreases with age. J Biol Chem. 1987;262:9948–51.
39. Talens RP, et al. Epigenetic variation during the adult lifespan: cross-sectional and longitudinal data on monozygotic twin pairs. Aging Cell. 2012;11:694–703. https://doi.org/10.1111/j.1474-9726.2012.00835.x.
40. Issa JP, Vertino PM, Boehm CD, Newsham IF, Baylin SB. Switch from monoallelic to biallelic human IGF2 promoter methylation during aging and carcinogenesis. Proc Natl Acad Sci U S A. 1996;93:11757–62.
41. Sakatani T, et al. Loss of imprinting of Igf2 alters intestinal maturation and tumorigenesis in mice. Science (New York, NY). 2005;307:1976–8. https://doi.org/10.1126/science.1108080.
42. Issa JP. Aging and epigenetic drift: a vicious cycle. J Clin Invest. 2014;124:24–9. https://doi.org/10.1172/jci69735.
43. Maekita T, et al. High levels of aberrant DNA methylation in Helicobacter pylori-infected gastric mucosae and its possible association with gastric cancer risk. Clin Cancer Res. 2006;12:989–95. https://doi.org/10.1158/1078-0432.ccr-05-2096.
44. Schulmann K, et al. Inactivation of p16, RUNX3, and HPP1 occurs early in Barrett's-associated neoplastic progression and predicts progression risk. Oncogene. 2005;24:4138–48. https://doi.org/10.1038/sj.onc.1208598.
45. Moss SF, Blaser MJ. Mechanisms of disease: inflammation and the origins of cancer. Nat Clin Pract Oncol. 2005;2:90–7. https://doi.org/10.1038/ncponc0081. quiz 91 p following 113.
46. Feil R, Fraga MF. Epigenetics and the environment: emerging patterns and implications. Nat Rev Genet. 2012;13:97–109. https://doi.org/10.1038/nrg3142.
47. Takahashi K, Yamanaka S. Induction of pluripotent stem cells from mouse embryonic and adult fibroblast cultures by defined factors. Cell. 2006;126:663–76. https://doi.org/10.1016/j.cell.2006.07.024.
48. Takahashi K, et al. Induction of pluripotent stem cells from adult human fibroblasts by defined factors. Cell. 2007;131:861–72. https://doi.org/10.1016/j.cell.2007.11.019.
49. Polo JM, et al. A molecular roadmap of reprogramming somatic cells into iPS cells. Cell. 2012;151:1617–32. https://doi.org/10.1016/j.cell.2012.11.039.
50. Chronis C, et al. Cooperative binding of transcription factors orchestrates reprogramming. Cell. 2017;168:442–459.e420. https://doi.org/10.1016/j.cell.2016.12.016.
51. Ben-Porath I, et al. An embryonic stem cell-like gene expression signature in poorly differentiated aggressive human tumors. Nat Genet. 2008;40:499–507. https://doi.org/10.1038/ng.127.

52. Gidekel S, Pizov G, Bergman Y, Pikarsky E. Oct-3/4 is a dose-dependent oncogenic fate determinant. Cancer Cell. 2003;4:361–70.
53. Hochedlinger K, Yamada Y, Beard C, Jaenisch R. Ectopic expression of Oct-4 blocks progenitor-cell differentiation and causes dysplasia in epithelial tissues. Cell. 2005;121:465–77. https://doi.org/10.1016/j.cell.2005.02.018.
54. Lu X, Mazur SJ, Lin T, Appella E, Xu Y. The pluripotency factor nanog promotes breast cancer tumorigenesis and metastasis. Oncogene. 2014;33:2655–64. https://doi.org/10.1038/onc.2013.209.
55. Boumahdi S, et al. SOX2 controls tumour initiation and cancer stem-cell functions in squamous-cell carcinoma. Nature. 2014;511:246–50. https://doi.org/10.1038/nature13305.
56. Blelloch RH, et al. Nuclear cloning of embryonal carcinoma cells. Proc Natl Acad Sci U S A. 2004;101:13985–90. https://doi.org/10.1073/pnas.0405015101.
57. Hochedlinger K, et al. Reprogramming of a melanoma genome by nuclear transplantation. Genes Dev. 2004;18:1875–85. https://doi.org/10.1101/gad.1213504.
58. Fodde R. The APC gene in colorectal cancer. Eur J Cancer. 2002;38:867–71.
59. Sanz MA, Martin G, Diaz-Mediavilla J. All-trans-retinoic acid in acute promyelocytic leukemia. N Engl J Med. 1998;338:393–4. https://doi.org/10.1056/nejm199802053380613.
60. Carette JE, et al. Generation of iPSCs from cultured human malignant cells. Blood. 2010;115:4039–42. https://doi.org/10.1182/blood-2009-07-231845.
61. Bernhardt M, et al. Melanoma-derived iPCCs show differential tumorigenicity and therapy response. Stem Cell Rep. 2017;8:1379–91. https://doi.org/10.1016/j.stemcr.2017.03.007.
62. Hashimoto K, et al. Cellular context-dependent consequences of Apc mutations on gene regulation and cellular behavior. Proc Natl Acad Sci U S A. 2017;114:758–63. https://doi.org/10.1073/pnas.1614197114.
63. Kim J, et al. An iPSC line from human pancreatic ductal adenocarcinoma undergoes early to invasive stages of pancreatic cancer progression. Cell Rep. 2013;3:2088–99. https://doi.org/10.1016/j.celrep.2013.05.036.
64. Suva ML, et al. Reconstructing and reprogramming the tumor-propagating potential of glioblastoma stem-like cells. Cell. 2014;157:580–94. https://doi.org/10.1016/j.cell.2014.02.030.
65. Shaffer SM, et al. Rare cell variability and drug-induced reprogramming as a mode of cancer drug resistance. Nature. 2017;546:431–5. https://doi.org/10.1038/nature22794.
66. Stadtfeld M, Maherali N, Borkent M, Hochedlinger K. A reprogrammable mouse strain from gene-targeted embryonic stem cells. Nat Methods. 2010;7:53–5. https://doi.org/10.1038/nmeth.1409.
67. Abad M, et al. Reprogramming in vivo produces teratomas and iPS cells with totipotency features. Nature. 2013;502:340–5. https://doi.org/10.1038/nature12586.
68. Ohnishi K, et al. Premature termination of reprogramming in vivo leads to cancer development through altered epigenetic regulation. Cell. 2014;156:663–77. https://doi.org/10.1016/j.cell.2014.01.005.
69. Ruteshouser EC, Robinson SM, Huff V. Wilms tumor genetics: mutations in WT1, WTX, and CTNNB1 account for only about one-third of tumors. Genes Chromosomes Cancer. 2008;47:461–70. https://doi.org/10.1002/gcc.20553.
70. Hong H, et al. Suppression of induced pluripotent stem cell generation by the p53-p21 pathway. Nature. 2009;460:1132–5. https://doi.org/10.1038/nature08235.
71. Li H, et al. The Ink4/Arf locus is a barrier for iPS cell reprogramming. Nature. 2009;460:1136–9. https://doi.org/10.1038/nature08290.
72. Liu XS, et al. Editing DNA methylation in the mammalian genome. Cell. 2016;167:233–247.e217. https://doi.org/10.1016/j.cell.2016.08.056.
73. Morita S, et al. Targeted DNA demethylation in vivo using dCas9-peptide repeat and scFv-TET1 catalytic domain fusions. Nat Biotechnol. 2016;34:1060–5. https://doi.org/10.1038/nbt.3658.

Chapter 4
Recapitulating Hematopoietic Development in a Dish

Kim Vanuytsel, Martin H. Steinberg, and George J. Murphy

Abstract The ability to generate patient-specific, induced pluripotent stem cells (iPSCs) together with the advent of gene-editing technologies has opened up a realm of opportunities for disease modeling, gene correction, and regenerative medicine applications aimed at better understanding and treating hematological disorders. The widespread use of reprogramming and gene-editing techniques has resulted in fine-tuning of these technologies to the point where they are an integral part of molecular biology research toolkits. The challenge remaining at this point is to achieve efficient and robust differentiation of pluripotent stem cells (PSCs) toward blood cells that resemble their in vivo counterparts. Here we provide an overview of our current understanding of in vivo hematopoietic development and how that has been used as a roadmap to guide in vitro hematopoietic development. We discuss recent advances and limitations encountered when recapitulating hematopoietic development in vitro. Finally, we highlight examples of how patient-specific iPSCs have been successfully used for the modeling of hematological disorders and how they have played a prominent role in uncovering pharmacologically targetable disease mechanisms.

Keywords Pluripotent stem cells · iPSCs · Hematopoietic stem cell · Hematopoiesis Erythropoiesis · Disease modeling · Hemoglobinopathies

---

K. Vanuytsel · M. H. Steinberg · G. J. Murphy (✉)
Section of Hematology and Oncology, Department of Medicine,
Boston University School of Medicine, Boston, MA, USA

Center for Regenerative Medicine (CReM), Boston University and Boston Medical Center,
Boston, MA, USA
e-mail: gjmurphy@bu.edu

H. Inoue, Y. Nakamura (eds.), *Medical Applications of iPS Cells*,
Current Human Cell Research and Applications,
https://doi.org/10.1007/978-981-13-3672-0_4

4.1 Introduction

The discovery that somatic cells can be reprogrammed into a pluripotent state has opened up tremendous opportunities for disease modeling and regenerative medicine. The ability to generate patient-specific, induced pluripotent stem cells (iPSCs) has enabled the study of a wide range of blood disorders in the exact genetic background of the patient. Parallel advances in gene-editing technology allow for the elimination of disease-causing mutations at the pluripotent stage, permitting the study of syngeneic, corrected lines and the rigorous validation of genetic modifications before the cells are differentiated into relevant cell types and potentially returned to the patient. Red blood cells and platelets, the affected terminally differentiated cells of hemoglobinopathies and thrombocytopenias, are excellent candidate cell types for such an approach as both are devoid of a nucleus, reducing the chance of downstream oncogenic transformation events due to genetic manipulation. Moreover, unmanipulated iPSC-derived erythrocytes and platelets have the potential to overcome the current shortage of donors as well as compatibility issues that arise with transfusion. Another very promising cell type for regenerative medicine is hematopoietic stem cells (HSCs) as they can functionally reconstitute the entire hematopoietic system of a recipient following transplantation [1]. This remarkable potential makes HSC transplantation an effective therapy for the treatment of hematological disorders [2] and puts the derivation of HSCs from iPSCs at the center of intense research efforts.

Critically, advances in disease modeling and potential regenerative medicine applications depend on how accurately the process of blood cell formation can be recapitulated in vitro as the resulting cells must match the characteristics and functional potential of naturally derived, postnatal blood cells. Significant progress has been made in elucidating the molecular signals that drive blood cell formation in the developing embryo. These developmental cues have subsequently been used as a roadmap to guide in vitro differentiation from pluripotent stem cells (PSCs). At the same time, studying hematopoietic ontogeny in vitro using human PSCs has provided insights into in vivo hematopoietic development, illustrating the value of PSCs for answering basic research questions, especially when limited accessibility of the earliest developmental stages in human hematopoietic development hampers the in-depth study of critical processes.

In this review, we will first discuss the current understanding of in vivo hematopoietic development. Next, the extent to which hematopoietic differentiation from PSCs recapitulates in vivo development will be discussed, including recent advances in the field. Here, challenges associated with the de novo generation of HSCs in culture and the determination of the developmental time frame of produced cells, as well as efforts to overcome current hurdles will be discussed. Lastly, the use of hPSCs as a surrogate system to study blood cell development in both health and hematological disease will be highlighted.

4.2 Mapping In Vivo Hematopoiesis to Direct In Vitro Blood Cell Development

Most of our knowledge about hematopoietic ontogeny has been obtained through the study of model organisms as the accessibility to the earliest developmental stages in these organisms far exceeds what can be achieved with human samples. Although confirmation in humans for many of these early hematopoietic events has lagged, it is now well established that human hematopoietic development follows a similar pattern as that seen in the mouse, despite differences in developmental timescales [3, 4]. Here, we offer an overview of the key stages of hematopoietic ontogeny as described for the mouse and refer to the human development (reviewed in [4]) where appropriate.

4.2.1 Primitive Wave (Yolk Sac) Hematopoietic Development

In the developing embryo, successive hematopoietic waves give rise to hematopoietic progenitors with increasing lineage potential (Fig. 4.1). A first wave of hematopoietic development, termed the primitive wave, arises in the yolk sac around embryonic day 7 (E7) in mice (16–18.5 days postconception (dpc) in human) where the extra-embryonic mesoderm gives rise to blood islands [4–7]. This transient wave produces a population of rapidly expanding primitive erythroid cells tasked with sustaining the oxygen needs of the rapidly developing embryo [6]. The primitive erythroid lineage is characterized by embryonic globin (βH1) expression. These cells enter the circulation as large nucleated erythrocytes, unlike their definitive counterparts, which are smaller and enucleate before entering circulation [7–9]. Notably, using reporter lines or antibodies that specifically label primitive cells, it has been demonstrated that primitive erythroid cells are capable of enucleating in circulation between E12.5 and E14.5, resulting in macrocytic enucleated erythrocytes [10, 11]. In addition to primitive erythrocytes, the primitive wave of hematopoiesis also gives rise to a population of macrophages and megakaryocytes [6, 12]. Although harder to distinguish from their definitive counterparts, primitive macrophages and megakaryocytes do display features unique to their embryonic origin. Compared with their adult counterparts, primitive megakaryocytes show reduced ploidy and lower platelet production [12, 13], and primitive macrophages appear to mature without passing through a monocyte intermediate stage [14, 15].

4.2.2 Erythro-myeloid Progenitor (EMP) Hematopoiesis

A day later, at approximately E8.25, a second wave of yolk sac-derived hematopoiesis produces definitive erythrocytes, megakaryocytes, and most myeloid lineages [6, 16]. This erythro-myeloid-restricted wave is referred to as erythro-myeloid

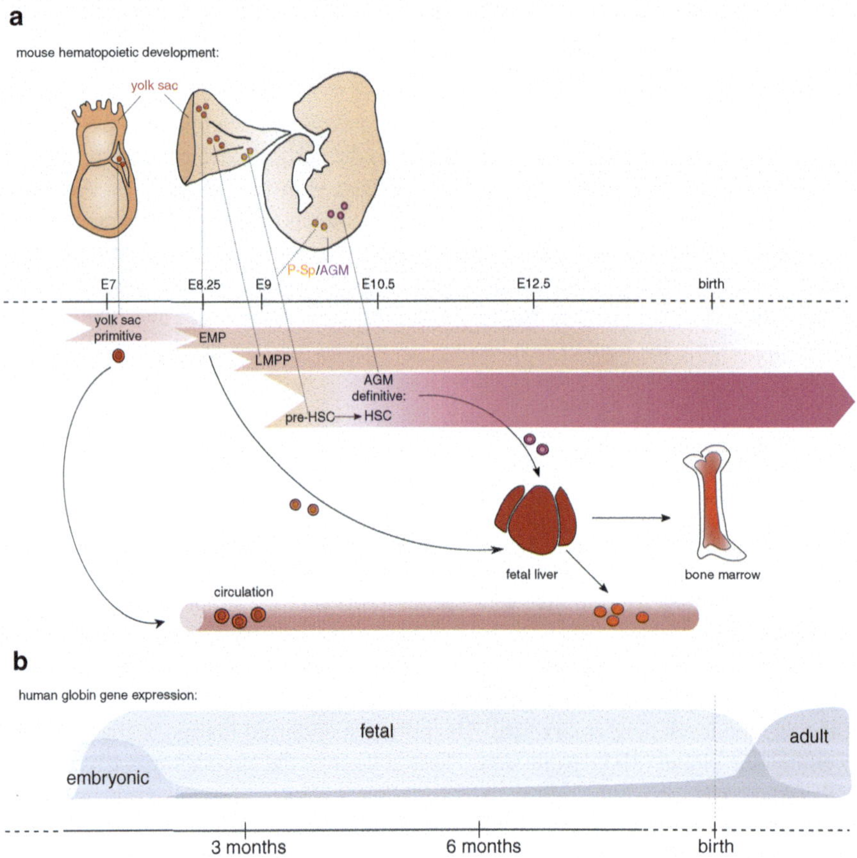

Fig. 4.1 Schematic overview of in vivo hematopoietic development. (**a**) Overview of the successive hematopoietic waves and progenitors generated during mouse development. (**b**) Overview of globin gene expression during human development. *EMP* erythro-myeloid progenitor, *LMPP* lymphoid-primed multipotent progenitor, *AGM* aorta-gonad-mesonephros, *P-Sp para*-aortic splanchnopleura, *HSC* hematopoietic stem cell

progenitor (EMP) hematopoiesis and is considered distinct from both primitive yolk sac and definitive aorta-gonad-mesonephros (AGM) hematopoiesis [8, 17]. This point is reflected in the unique globin expression profile of EMP-derived erythroid cells, where low levels of embryonic βH1 globin together with adult (β1) globin generate a pattern that differs from both primitive and AGM-derived definitive erythroid cells [18]. Although this wave has been well characterized in the mouse and attempts to extrapolate these findings to the human situation have been made by incorporating a human transgenic β-globin locus into mice [6, 8, 18], EMP hematopoiesis has not been formally characterized during human development [4]. Nevertheless, burst-forming unit-erythroid (BFU-E) cells, which are considered the first definitive erythroid progenitors arising during hematopoietic ontogeny, are first detected in the human yolk sac around 28–35 dpc [19]. This similarity to their first appearance in mice, which coincides with the onset of EMP hematopoiesis, strongly suggests the emergence of a similar EMP wave in humans at about 4–5 weeks in gestation [4, 8].

4.2.3 Lymphoid-Primed Multipotent Progenitor (LMPP) Hematopoiesis

At approximately E9.0, B and T cell potential can be detected in the yolk sac as well as in the embryonic para-aortic splanchnopleura (P-Sp) [20, 21]. Emerging independently from definitive HSCs in the AGM, this early wave also displays myeloid potential in addition to lymphoid potential, hence the designation LMPP hematopoiesis [22]. This initial program of B cell lymphopoiesis shows a bias toward production of innate type B-1 and marginal zone B cells instead of B-2 cells, which are considered to be generated through a separate HSC-dependent wave [20, 23]. In terms of T cell potential, both fetal and adult-type T cells (αβ and γδ subsets) are produced by the YS- and P-Sp-derived T progenitor cells [21]. LMMPs therefore might present a source of lymphoid cells with a yolk sac origin that persist into adulthood [20, 21]. Similarly, tissue-resident macrophages found in the lungs and liver appear to originate from the early waves of hematopoiesis in the yolk sac, as well as the microglia of the brain [24–27].

4.2.4 Definitive Wave (AGM) Hematopoietic Development

The definitive wave of hematopoiesis is characterized by its ability to give rise to hematopoietic stem cells (HSCs) that exhibit long-term, multi-lineage engraftment potential. Cells corresponding to this definition first emerge at the ventral wall of the dorsal aorta within the AGM region at E10.5 in the mouse and at 32dpc in human embryos [28–31] but can also be detected in other major arteries of the developing embryo such as the vitelline and umbilical arteries [32]. By E11.5, HSCs are also present in the yolk sac and the placenta [33, 34]. As the heart has already started beating at this time point, creating circulation of blood cells throughout the conceptus, it is difficult to decipher whether these sites should be regarded as de novo generators of HSCs or instead contributors to the HSC pool via expansion of HSCs generated elsewhere.

Via the circulation, hematopoietic progenitors generated at different developmental sites inside or outside of the embryo home to the fetal liver, where they undergo massive expansion before being distributed throughout the bone marrow, thymus, and spleen [35]. The transition from the fetal liver to the bone marrow as the main site of hematopoiesis occurs at approximately E17.5 in mouse or

10–11 weeks in human embryos and is not complete until after birth [36, 37]. The pool of HSCs arising prenatally through expansion in the fetal liver forms the basis for life-long blood production in adults [1, 35, 38]. However, as mentioned earlier, certain subsets of lymphoid cells as well as tissue-resident macrophages found in the lung, liver, and brain postnatally are believed to originate from the early waves of hematopoiesis in the yolk sac through HSC-independent mechanisms [20, 21, 24–27].

4.2.4.1 Hemogenic Endothelium and Endothelial-to-Hematopoietic Transition

The spatial localization and marker expression profile of the first definitive adult-repopulating HSCs suggest an endothelial origin for progenitors of the definitive wave. HSCs appear as clusters tightly adhered to the ventral endothelium of the dorsal aorta in the AGM region and express a range of markers shared with endothelial cells in this location such as Sca-1, Runx1, c-kit, VE-cadherin, CD34, SCL, and Gata2 [31, 38–42]. During this process, referred to as endothelial-to-hematopoietic transition (EHT), hematopoietic clusters bud off into the lumen of the aorta and start expressing CD45, which distinguishes the newly formed HSCs from the adjacent endothelium [38, 41, 43]. The cell population giving rise to the first definitive HSCs is termed hemogenic endothelium, and besides HSCs in the AGM [41], also E8.25 yolk sac EMPs [44] and E9.5 B and T cell progenitors [20, 21] have been shown to emerge from a hemogenic endothelial cell population. Interestingly, an early extra-embryonic mesodermal origin has been suggested for all hemogenic endothelial cells as endothelial progenitors emerging in the yolks sac around E7–7.5 have been found to migrate to intra-embryonic sites where they can contribute to hemogenic endothelium in the dorsal aorta and even HSCs [45–48]. Finally, and importantly, it is clear that all blood cells emerge from cells with an endothelial signature through an EHT stage, and this process seems to be conserved in vertebrate hematopoiesis as it has been described in different species including frogs, chicks, mice, and humans [49–55].

4.2.4.2 Origins of the Hematopoietic Stem Cell

At E9, prior the emergence of HSCs with adult-repopulating potential, both the yolk sac and the P-Sp (pre-AGM region) produce progenitors that have the capacity to engraft neonatal but not adult recipients [56, 57]. However, the marrow isolated from these primary recipients is able to engraft secondary adult recipients, indicating that an intermediate in vivo maturational stage can endow the progeny of these progenitors with adult-repopulating potential [57]. This suggests that these progenitors represent immature HSCs that need to undergo further maturational patterning before they display their true HSC identity. Similarly, following an in vitro maturation step using OP9 stromal cells or endothelial cells in a co-culture setting,

E9.5–E10 AGM progenitors that are not capable of engrafting adult recipients can become competent, engraftable HSCs [58–60]. The fact that these early progenitors eventually acquire the potential to generate HSCs has led to their designation as HSC precursors or pre-HSCs. It is interesting to note how, leading up to the emergence of true definitive HSCs, progenitors with increasing lineage potential are generated during successive waves of development, whereas in adult hematopoiesis, HSCs reside at the apex of a hierarchical system that generates downstream progenitors displaying more and more restricted lineage potential.

4.2.5 *Identification and Characterization of Primitive Versus Definitive Wave Hematopoiesis*

Due to the different spatial and temporal emergence of progenitors with increasing hematopoietic potential, the need arose to distinguish different programs in development. This led to the initial distinction between a primitive and definitive wave of hematopoiesis when it was discovered that hematopoiesis originates independently at two locations, in the yolk sac and in the AGM [17, 61]. More recently, studies describing EMPs and LMPPs within the yolk sac have added to the complexity of hematopoietic ontology [18, 20–22]. This has only amplified the confusion in terms of nomenclature as in the past progenitors from both of these waves have been referred to as both primitive and definitive [62–65], whereas it is not clear whether they should be assigned to either of the proposed categories [4, 17]. T cell potential, for instance, has been associated with acquisition of a "definitive" status [64, 66]. However, as B and T cell progenitors arise in the yolk sac before the emergence of definitive AGM hematopoiesis [20, 21, 67], designation of lymphoid potential as a definitive characteristic is dependent on context and should be interpreted with caution as it is not necessarily synonymous with AGM definitive status and therefore does not automatically refer to a program with true HSC competence.

4.2.5.1 Using Globin Expression to Identify Subsequent Hematopoietic Programs

Concerning the erythroid lineage, there seems to be a consensus to refer to erythroid cells produced during the initial hematopoietic wave in the YS as primitive and everything afterward as definitive. Under this definition, the onset of definitive erythropoiesis coincides with the emergence of BFU-Es in the yolk sac, which in turn is associated spatially and temporally with the emergence of EMPs [8].

Depending on the developmental program, different genes from the β-globin gene cluster are transcribed in erythroid cells, resulting in a unique globin expression pattern that can be used to define the different programs. The first erythroid population that arises in the yolk sac consists of relatively large, nucleated primitive erythroid cells, predominantly expressing embryonic globin [7, 8, 68]. In mice this means that εγ- and βH1-globin genes from the beta globin cluster are expressed, and in the human situation, this translates to predominant embryonic (ε)-globin expression as well as some fetal (γ)-globin expression. Primitive erythroblasts from both species also express ζ-globin and α-globin from the α-globin gene cluster, and the combination of these different globin chains from the α- and β-globin cluster leads to a variety of hemoglobin tetramers [8, 68].

This initial wave of primitive erythropoiesis is followed by definitive erythropoiesis, which has two distinct developmental origins [8]. EMPs in the yolk sac, and, slightly later, HSCs in the AGM can both give rise to progenitors of the definitive erythroid program. These progenitors then travel to the fetal liver where they differentiate to produce the first definitive erythrocytes. The smaller cell size and the fact that definitive erythrocytes expel their nucleus before entering circulation allows for a straightforward distinction from primitive erythrocytes. Discriminating between both types of cells arising from the definitive erythroid program is a more challenging task. Nevertheless, careful study of EMPs in the mouse has uncovered a unique globin expression profile of EMP-derived erythroid cells where low levels of embryonic (βH1) globin in combination with adult (β1) globin expression generate a pattern that differs from both primitive and AGM-derived definitive erythroid cells [18]. EMPs have not been formally characterized during human development [4]. However, when the same EMP-derived erythroid population in mice was assessed in the context of a transgenic human β-globin locus, again a unique globin expression signature emerged, showing co-expression of human embryonic (ε), fetal (γ), and adult (β) globins, distinguishing EMP-derived erythroid cells from AGM-derived definitive erythroid cells on the basis of embryonic globin expression in the former [18].

Fetal (γ)-globin expression, a feature unique to anthropoid primates, gets silenced postnatally when globin expression switches to adult β-globin and the bone marrow (spleen in mice) becomes the major sites of erythropoiesis [68] (Fig. 4.1b). This means that β-globin chains replace the γ-globin chains in hemoglobin tetramers resulting in a switch from HbF ($\alpha 2\gamma 2$) to HbA1 ($\alpha 2\beta 2$) with HbA1 accounting for 97% of all hemoglobin in adult erythrocytes. The other percentages are shared between HbA2 ($\alpha 2\delta 2$) (2%) and HbF (1%) [68].

4.3 In Vitro Hematopoiesis Through the Lens of a Pluripotent Stem Cell Model

In vitro hematopoietic differentiation protocols are designed to mimic the developmental processes that take place in the early embryo [69, 70]. First established using mouse PSCs, in vitro differentiation protocols were translated to human PSCs, showing a similar progression through the different developmental stages. To induce hematopoietic differentiation from PSCs, several approaches have been explored, such as stromal cell co-culture systems [71–78], embryoid body (EB) formation [69, 79–81], and monolayer culture systems [82–84]. Whereas stromal co-culture systems generally depend on serum-containing medium, other approaches can be executed in fully defined conditions aimed at reducing variability and excluding unknown factors present in serum that preclude detailed analysis of critical signaling pathways. Irrespective of the system used, kinetic analysis has revealed that differentiation cultures follow a similar path toward hematopoietic commitment. Upon induction of differentiation, PSCs pass through a primitive streak stage, where markers such as BRACHYURY and MIXL1 are transiently upregulated. Next, differentiating cells progress to a mesodermal stage, and expression of markers such as KDR (Flk-1), RUNX1, and CD34 can be detected [70, 80, 81, 85–87]. Upon further differentiation, markers indicative of hematopoietic commitment such as CD43 and CD45 get expressed on the cell surface during a process resembling endothelial-to-hematopoietic transition (EHT) [72, 73, 88, 89]. This sequence is followed spontaneously when differentiation of hPSCs is induced through EB formation in the presence of serum, even in the absence of growth factors, and recapitulates early hematopoietic development as it occurs in vivo [69]. Most often however, signals that guide hematopoietic development in the embryo are mimicked through the sequential addition of different cytokine and growth factor cocktails to increase the efficiency of specification.

Studies using hPSCs have verified that, as in model organisms, PS and mesoderm induction during in vitro hematopoietic differentiation is regulated by BMP4, bFGF, Activin/Nodal, and WNT/b-catenin signaling [79, 81, 87, 90]. The combination of BMP4, a ventral mesoderm inducer, and a spectrum of hematopoietic cytokines was found to strongly promote hematopoietic differentiation from hESC [79]. In agreement with this finding, another study illustrated that addition of BMP4 to a serum-free, feeder-free EB differentiation system could induce the expression of genes marking hematopoietic mesoderm and support the emergence of hematopoietic progenitor cells. In this study, VEGF appeared to be a crucial companion for the generation of robust numbers of hematopoietic colony-forming cells and more mature descendants. Addition of SCF and bFGF to this cocktail further increased the total yield of hematopoietic cells generated from hESCs [87]. Once hematopoietic mesoderm is formed and EHT is initiated, further specification of the emerging hematopoietic progenitors can be guided through addition of lineage-specific cytokines and growth factors. Erythropoietin (EPO), for instance, is the main regulator of erythropoiesis [91], whereas Notch is a crucial mediator of T cell development [92, 93]. Thrombopoietin (TPO), on the other hand, fulfills a dual role as this growth factor is the primary stimulus for the megakaryocytic lineage but is also involved in the maintenance of HSC in a quiescent state [94, 95].

4.3.1 In Vitro Recapitulation of Successive Hematopoietic Programs

As in natural development, the successive waves of hematopoiesis must be recapitulated during in vitro differentiation from PSCs. This process can occur spontaneously, in the absence of specific growth factors, as well as during directed differentiation approaches [66, 69, 96, 97]. In contrast to in vivo specification, which is impacted by a dynamic niche, there is no spatiotemporal separation of the different hematopoietic programs during in vitro differentiation from PSCs. Most reports in the literature describe the onset of a hematopoietic program reflecting the primitive wave of hematopoiesis, often through a hemangioblast-like intermediate, before the appearance of hematopoietic cells with more definitive characteristics [66, 69, 97, 98]. However, as subsequent transcriptional programs and waves of development emerge within the same culture dish and overlap in time, distinguishing between them is challenging. With respect to this point, CD235a expression has been described to mark a subset of mesodermal cells that will give rise to a primitive population of cells, whereas exclusion of cells with this marker leads to the development of more definitively patterned cells from PSCs [66]. Similarly, a distinction between different developmental programs can be made based on signaling pathways important for their specification early on in mesoderm formation. Whereas the primitive program relies on Activin/Nodal signaling between days 2 and 4 of differentiation, a more definitive wave of hematopoietic progenitors can be obtained by inhibiting this pathway and/or stimulation of Wnt signaling during the same time frame [64]. Lymphocyte potential and Notch dependence have been used in these studies to define the onset of the definitive program; however as T and B cell potential can arise in the yolk sac before the onset of the definitive AGM hematopoietic program [20, 21, 67], this does not necessarily indicate an intra-embryonic origin or the capacity to generate true HSC-like cells from this more definitive program. Nevertheless, the finding that manipulation of Wnt and Activin/Nodal signaling early in the differentiation leads to increased expression of the *HOXA* gene cluster in the resulting $CD34^{+}$ hematopoietic progenitors and results in a gene expression pattern for PSC-derived cells that resembles that of human AGM-derived cells argues in favor of an AGM origin [99]. However, in the absence of human yolk sac-derived progenitors for comparison, it remains unclear whether these patterning changes increase the resemblance to the AGM definitive program at the cost of a YS identity or whether they increase the similarity to "more definitive" progenitors found at both locations equally while moving further away from the primitive program. Given that subtle differences in *HOXA* expression remain compared to the human AGM samples and that the hPSC-derived progenitors still lack essential characteristics associated with long-term hematopoietic reconstitution [99], these HOXA+ progenitors could possibly also represent pre-HSC which are detected both in the YS and P-Sp (pre-AGM region) prior to definitive AGM hematopoiesis. More research will be necessary to elucidate the exact staging in development of PSC-derived cells; however it is interesting to note that interfering with signaling pathways during PS formation can have a distinct effect on hematopoietic progenitors arising much later in the differentiation. These findings suggest that the distinction between extra- and intra-embryonic hematopoietic fates or at least between different hematopoietic programs is made at the time of mesodermal patterning [4].

4.3.2 *Endothelial-to-Hematopoietic Transition (EHT) in the Dish*

Development of blast colony-forming cells (BL-CFCs) that can give rise to both vascular and hematopoietic progeny early in the hematopoietic differentiation from hPSCs has been shown to precede the establishment of the primitive hematopoietic program [81, 86, 100]. Interestingly, the first functional evidence for the existence of such a bipotential progenitor came from mouse PSC differentiation cultures [101, 102] and was subsequently confirmed in vivo [103]. This is an excellent example of how studying hematopoietic differentiation from PSCs in vitro can yield insights into in vivo development and argues for the use of this surrogate system to study early developmental processes that might otherwise not be easily accessible. The hemogenic endothelium that gives rise to subsequent waves of hematopoiesis has also been extensively characterized in PSC-derived cultures and has been shown to express RUNX1 and GFI1 [97, 99, 104, 105] but not CD184 (CXCR4), DLL4, or CD73 [97, 106], separating it from venous and arterial vascular endothelium generated from PSCs [106]. Currently, no surface markers are available that can distinguish between hemogenic endothelium that gives rise to definitive adult-repopulating HSCs and hemogenic endothelium that generates hematopoietic progenitors with more limited developmental and engraftment potential.

Together, these studies indicate that all blood cells emerge from cells with an endothelial signature through EHT and that this transition can be considered a common mechanism underlying blood formation during both in vivo development and in vitro hematopoietic differentiation from PSCs.

4.3.3 *Engineering Hematopoietic Stem Cells from Pluripotent Stem Cells*

Although the field has made significant progress by harnessing developmental signals found in the embryo to guide in vitro hematopoietic development from PSCs, challenges still remain. The exact patterning required for de novo HSC generation remains largely unknown. As HSCs represent the most potent hematopoietic cells, capable of reconstituting the entire hematopoietic system in a transplantation setting, their generation from PSCs has been at the center of the research efforts of many groups. What makes this quest particularly challenging is that even though HSCs can be characterized and enriched based on molecular criteria such as the expression of certain cell-surface markers, true HSC identity can only be fully confirmed based on the fulfillment of functional criteria [1]. By definition, HSCs are capable of long-term self-renewal and multi-lineage reconstitution, which means that their derivation can only be confirmed following transplantation when introduced cells give rise to all hematopoietic lineages in the recipient for an extended period of time [107]. Due to ethical and practical reasons, a xenotransplantation setting is inevitable when testing the potential of hPSC-derived progenitors. Several immunocompromised mouse models have been developed over the years aimed at providing an environment that is supportive for the engraftment of human cells [108]. Although this has been relatively successful for human postnatal HSCs with cord blood-derived cells showing engraftment levels up to 40% [109, 110], the levels of engraftment reported from hPSC-derived progenitors have been exceptionally low and mostly restricted to the myeloid lineages [76, 111, 112]. At this point, it is challenging to decipher what could be the reason for the low engraftment from hPSC-derived cells as there are many possible contributing factors. Besides the obvious hurdle of relying on a xenograft setting for a read-out of HSC potential, the vaguely defined developmental time frame of hPSC-derived cells is also a confounding factor. Whether the most advanced hematopoietic progenitors generated in vitro from PSCs represent a more definitive yolk sac stage of development or alternatively AGM-type definitive progenitors is still under debate as there are currently no convincing markers to unequivocally distinguish both types of progenitors. If we consider the scenario where in vitro hematopoietic differentiation protocols from PSCs are mostly geared toward the recapitulation of a yolk sac hematopoietic program with a transient primitive wave followed by a more definitive type wave, then the patterning early on might need to be revised to arrive at an intra-embryonic hemogenic endothelium population that can give rise to true adult-repopulating HSCs. This idea is currently being explored by several groups focusing on manipulation of key pathways during the early specification of mesoderm and subsequent characterization of the resulting hematopoietic progenitors [64, 99].

4.3.3.1 In Vivo Differentiation Through Teratoma Formation: The Mouse as Incubator

In an attempt to overcome the incomplete understanding of the signaling cascade that leads to the correct patterning and/or maturation of HSCs in vitro, two groups bridged this knowledge gap using an in vivo teratoma setting [113, 114]. When hPSCs are injected ectopically in immunocompromised mice, they have the capacity to form all three germ layers, guided by the inductive environment found in the mouse. During this process that results in a mass of disorganized tissues and organs, called a teratoma, bone marrow elements can be formed [114, 115]. Making use of endogenous specification signals, both groups found that it was possible to derive engraftable $CD34^{+}CD45^{+}$ progenitors using this technique. Even at the very low efficiencies noted in these works, demonstration of donor-derived multilineage peripheral blood chimerism is an encouraging finding that indicates that transplantable hematopoietic progenitors can indeed be generated from hPSCs in the proper setting. Nevertheless, it remains to be demonstrated whether these progenitors represent bona fide HSCs as the time frame used to assess hematopoietic reconstitution was not sufficiently long to confirm true long-term hematopoietic potential [30, 107]. In these studies, the in vivo specification of hPSCs toward HSC-like cells was aided by co-injection of OP9 stromal cells and/or continuous administration of hematopoietic cytokines. In a more recent report, these supportive measures were replaced by overexpression of a set of transcription factors, significantly speeding up the process of HSC generation from mPSC [116].

4.3.3.2 Induction of an HSC Program Through Forced Expression of Exogenous Factors

The introduction of transcription factors to boost the in vitro generation of HSCs illustrates an alternative approach to overcome challenges associated with the directed differentiation of PSCs to HSCs. The demonstration that fibroblasts can be reprogrammed to iPSCs via introduction of four transcription factors roughly a decade ago was a powerful example of the fact that cellular identity can be manipulated by enforcing gene expression of the right combination of factors [117]. Since then, this transgene overexpression strategy has been employed to either direct the in vitro differentiation of PSCs or to transdifferentiate somatic cell types, a process also referred to as direct lineage conversion. Different cell types have been targeted this way, all attempting to generate HSC-like cells, including PSCs [118, 119], fibroblasts [120, 121], hematopoietic cells [122, 123], and endothelial cells [124, 125]. One of the most recent reports documents the generation of human HSC-like cells using the forced expression of seven factors (*ERG*, *HOXA5*, *HOXA9*, *HOXA10*, *LCOR*, *RUNX1*, and *SPI1*) to confer HSC-like engraftment, self-renewal, and multilineage capacity onto hPSC-derived hemogenic endothelial cells [126]. Although this engineered induction of HSC-like characteristics positions the converted cells within the functional definition of HSCs, the authors note limitations in the robustness of engraftment and the recapitulation of terminal differentiation compared to bona fide HSCs. Nevertheless, the generation of cells that are tantalizingly close to bona fide HSCs from hPSCs advances the field in the direction of realizing the enormous potential that PSCs hold in terms of regenerative medicine. Although the presence of exogenous factors prevents a straightforward translation of the resulting HSC-like cells into a clinical setting at this point, further studying the factors that enable this conversion is expected to unravel pathways and circuits involved in HSC specification during development. This will instruct directed differentiation approaches aimed at achieving a similar fate change through manipulation of relevant pathways in a sequential manner using small molecules and growth factors rather than transgenes.

4.3.4 *Initiation of Globin Switching in PSC-Derived Erythroid-Lineage Cells*

During human erythroid cell development, a maturational switch occurs from ζ-(*HBZ*) to α-globin (*HBA*) gene expression and from embryonic (ε; *HBE*) to fetal (γ; *HBG*) gene expression during the first trimester of conception. This is followed by a second switch from fetal (γ; *HBG*) to adult (β; *HBB*) globin gene expression beginning in mid-gestation that is largely completed by 6 months after birth [127–130]. Notably, one can speculate that this switch might not occur in all cells simultaneously with the timing of this change likely to be genetically controlled as it varies among individuals. Hemoglobin analysis shows that the first two globin switches (ζ to α and ε to γ) are recapitulated during in vitro differentiation, again demonstrating the usefulness of hPSCs to model early hematopoietic development [77, 99]. The γ- to β-switch, however, has been more difficult to capture in PSC-derived progeny. Although several reports describe induction of β-globin expression in hPSC-derived erythroid cells, none of these platforms demonstrate a complete switch from predominantly γ-globin expressing cells to predominantly β-globin expressing cells as occurs in vivo [77, 78, 131]. This is not completely unexpected as this switch is only completed postnatally and PSC-derived cells typically have a more immature/prenatal character [132]. Nevertheless, the observed levels of generated β-globin are encouraging and provide information on the developmental positioning of differentiated cells as β-globin expression is a characteristic that distinguishes the definitive from the primitive erythroid program. Whether in vitro differentiated β-globin expressing cells are more akin to EMP-derived erythroid cells or definitive AGM-derived erythroid cells depends on the globin co-expression pattern displayed by the cells. This is typically difficult to decipher in bulk cultures where several hematopoietic programs can coexist. For this reason, techniques that allow for the assessment of globin co-expression at the single cell level will be most insightful [133].

4.4 Harnessing iPSCS for Hematological Disease Modeling and Drug Discovery

Sampling primary hematopoietic tissues for disease modeling and drug discovery can provide only limited numbers of cells with limited proliferative potential. In addition, the tissue type of interest could be hard to obtain because it is destroyed by the disease process or because invasive procedures are required to isolate the relevant cells. This is often the case for bone marrow failure syndromes where bone marrow aspiration might not yield sufficient cells, especially in pediatric patients where these disorders commonly present [134]. Rare disorders present another challenge, as it is often impossible to recruit sufficient patients to get statistically relevant numbers of primary samples. In situations like these, renewable patient-specific cell sources such as iPSCs offer a powerful platform that allows for repeated testing of hypotheses, possibly after genetic manipulation or addition of certain compounds that could interfere with the disease process. Moreover, differentiation from iPSCs encompasses developmental stages that do not exist beyond the embryo, and for this reason, iPSCs present an invaluable tool to study disorders where the hematopoietic defect is thought to arise prenatally, as in Fanconi anemia [135, 136]. Finally, by using iPSCs, it is also possible to study multiple cell types derived from the same genetic background without the need for repeated biopsies making this a versatile platform to study multi-systemic blood disorders such as β-thalassemia where the interplay between inefficient erythropoiesis and iron homeostasis in the liver perpetuates the disease phenotype [137–139].

4.4.1 Hematological Disorders Affecting Hematopoietic Stem and Progenitor Cells (HSPCs)

A wide variety of iPSC lines generated from patients with blood disorders exists, including dyskeratosis congenita [140], Fanconi anemia [141], Shwachman-Diamond syndrome [142], Pearson syndrome [143], Diamond-Blackfan anemia [144], sickle cell anemia [145–147], β-thalassemia [148–151], imatinib-sensitive chronic myelogenous leukemia (CML) [152], and others. To accurately model such disorders and see the full extent of the disease-specific phenotype reveal itself, iPSCs will need to be differentiated into the target cell type that is affected. Some of these conditions represent bone marrow failure syndromes, affecting all three blood lineages and are therefore considered to arise from defects in the HSC compartment. Despite some of the challenges associated with generating true HSCs in vitro, PSC modeling of hematopoietic disorders has been successful at elucidating certain disease-specific phenotypes, especially if downstream progenitors are affected. Disease modeling using iPSCs derived from Shwachman-Diamond syndrome patients, for instance, has shown that protease-mediated autodigestion could be considered a common mechanism underlying both hematopoietic and pancreatic manifestations of the disorder and identified aprotinin as a potential therapeutic compound [142]. A recent report on PSC modeling of dyskeratosis congenita could even demonstrate a differential impact of disease-specific telomere shortening on progenitors from different hematopoietic programs, showing that primitive hematopoiesis was relatively unaffected compared with progenitors from a more definitive Wnt-stimulated program [153]. Modeling of Fanconi anemia (FA) using PSCs has also yielded valuable insights into both the hematopoietic defects associated with the disorder and the importance of the Fanconi anemia pathway in the context of pluripotency. As FA cells were shown to be resistant [154], though not refractory [141], to reprogramming due to the inherent defect in DNA repair, the first proof of principle of using PSCs to model FA was reported using a stable knockdown approach in hESCs [136]. Here, significant reductions in hematopoietic gene expression and progenitor numbers were noted upon in vitro differentiation of hPSCs with a stable knockdown of FA genes into hematopoietic progenitors. Once some of the hurdles limiting reprogramming efficiency were overcome [141], accounts of the resulting iPSCs [155, 156], as well as FANCA knockout hESCs [157], shed light on the importance of a functional FA pathway for proliferation and maintenance of a normal karyotype in pluripotent cells. Complementation of FA-deficient PSCs with a transgenic *FANCA* copy has been used to demonstrate restoration of the hematopoietic capacity of the cells, highlighting the use of FA-specific PSCs to model the disorder and screen for approaches that can improve the hematopoietic phenotype [136, 141]. This was further corroborated by a report using FA-specific iPSCs as a drug screening platform, confirming the action of compounds known to improve FA phenotypes and even identifying new compounds with a beneficial effect on the hematopoietic phenotype of both FA-specific hPSCs and bone marrow cells from FA patients [158].

4.4.2 Hemoglobinopathies: From Gene Correction and Disease Modeling to Drug Discovery and Validation in a Patient-Specific Genetic Context

The β-hemoglobinopathies are red blood cell disorders in which adult hemoglobin (HbA) is affected by mutations in the β-globin gene (*HBB*), resulting in anemia shortly after birth as globin expression switches from fetal to adult globin [159, 160]. Patient-specific iPSCs from both β-thalassemia [148–151] and sickle cell disease [145–147] have been used for the demonstration of proof of principle for the correction of the genetic defect in *HBB*. In these reports, the *HBB* gene is corrected at the pluripotent stage, even if the underlying disease-causing mutation is not expressed in PSCs, and the cells are subsequently differentiated toward the erythroid lineage to demonstrate that correction of the disease-causing mutation results in wildtype β-globin transcripts [145, 151] or protein [148–150]. This process illustrates the potential of gene correction and subsequent transfusion of the corrected cells as cell therapy for patients suffering from β-hemoglobinopathies where one functional copy of *HBB* is sufficient to alleviate the burden of the disease. Using these hemoglobinopathy-specific iPSCs to study the hematopoietic phenotype however is thwarted due to the limited expression of β-globin in PSC-derived erythroid cells. As a consequence of the prenatal character of PSC-derived cells, in vitro differentiated erythroid cells predominantly express embryonic and/or fetal globins [78, 99, 161]. As high levels of fetal hemoglobin (HbF) are ameliorative for these conditions and prevent effects such as the sickling of red blood cells in sickle cell anemia and compensate for the reduction or absence of HbA in β-thalassemia [162–165], these hemoglobinopathies will only lend themselves to disease modeling once β-globin expression can be sufficiently induced in PSC-derived erythroid cells. Reports demonstrating a switch to β-globin expression upon maturation of PSC-derived hematopoietic progenitors in mice [113, 122] are encouraging and offer the prospect of recapitulating this maturational step in vitro once the signals involved are fully elucidated. Moreover, efforts are being made to overcome the poor recapitulation of erythropoiesis in xenograft mouse models [166–170] with the aim of using such model systems to study the disease-specific phenotype of red blood cell disorders in an in vivo setting.

Perhaps unique to the β-hemoglobinopathies, an iPSC-derived red blood cell precursor that fails to switch from fetal to adult globin synthesis would be a useful cellular therapeutic. Predominantly HbF-containing erythrocytes are adequate oxygen transporters, and high amounts of HbF will prevent polymerization of any deoxy sickle hemoglobin that might be synthesized [171]. Moreover, isogenic cells are unlikely to evoke an immune response. Thus, without manipulation of the genome or chronic immunosuppression, a potentially curative cellular therapy that could be transfused in any infusion center could become widely available.

While human iPSC-derived erythroid cells in vitro do not undergo the switch from fetal (*HBG*) to adult (*HBB*) globin gene expression, they might prove to be a valuable resource for the identification of novel HbF inducers for the treatment of β-hemoglobinopathies. Currently, prior to clinical trials, in vitro drug testing is typically performed using immortalized human cells. However, such cell lines have undergone genetic alterations to ensure their immortalization, possibly altering the fidelity of the drug screens. The use of immortalized cell lines is a common cause of high attrition rates for drug development, as what works in vitro in one genetic background, and subsequently in animal models, does not always translate to the clinic [172]. As there are many different biological and demographic factors that might affect a hemoglobinopathy patient's response to a drug such as the presence of quantitative trait loci (QTL) effecting globin gene expression, ethnicity, age, and pretreatment HbF levels, exploring the effect of each is important. iPSCs, due to their unlimited supply of material and their ability to capture the exact genetic background of patients, represent a powerful platform for drug screening with the ability to correlate genetic variance to drug response (Fig. 4.2). The creation of an ethnically diverse library of sickle cell disease-specific iPSCs from three distinct geographic locations to represent multiple QTL and disease phenotypes [146], for instance, is expected to facilitate high-throughput screening of novel therapeutics in a manner that allows for a comprehensive analysis of patients and their individual responses, driving more individualized treatment plans for sickle cell disease patients. Moreover, utilization of iPSCs for drug screening presents researchers with the opportunity to explore the effects of a drug on multiple cell types from the same individual. Common side effects seen during drug testing are due to toxicity that can affect the peripheral nervous system, liver, and heart, making examination of those systems crucial for drug screening in vitro. Given the pluripotent nature of iPSCs, it is possible to obtain toxicity and efficacy data in multiple cell types with the same genetic background during a complete "*in vitro* clinical trial" prior to in vivo studies [173–175]. The ability to use iPSCs for in vitro disease modeling, for high-throughput drug screening, and for drug toxicity screening could ultimately help to reduce the time that elapses from the identification of novel therapeutics to FDA approval.

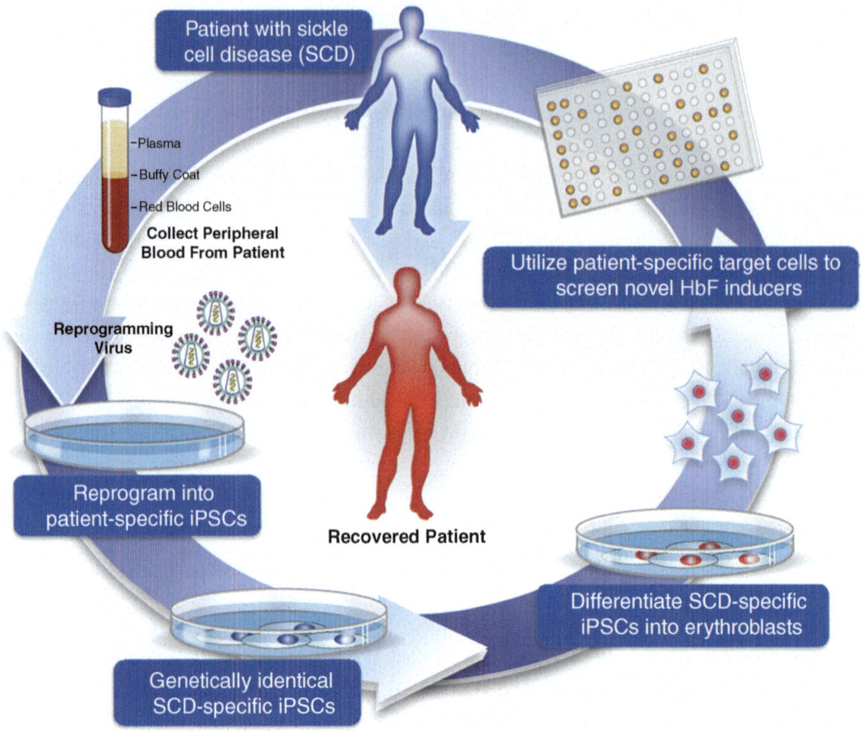

Fig. 4.2 iPSCs generated from peripheral blood of sickle cell disease patients can be differentiated into the cell type affected by the disorder and subsequently used to screen novel therapeutics in a high-throughput manner

4.5 Conclusion

iPSCs represent a versatile platform for the study of developmental processes both in health and disease as well as for drug screening in the context of individual patients and their unique genetic background. Proof-of-principle studies have demonstrated the feasibility of genetic correction for hematological disorders and differentiation of the corrected cells into the relevant cell types with the prospect of using gene-corrected blood cells in regenerative medicine approaches. Techniques such as reprogramming and gene-editing technology that have revolutionized this field of research have matured to the point where they are routine practice. At this point, the biggest remaining challenge is the efficient and robust differentiation of hPSC toward blood cells that resemble their in vivo counterparts in terms of physical properties and engraftment potential. Engineering efforts to equip cells with the desired characteristics, through forced expression of transcription factors or with the help of a mouse in vivo setting, have brought us tantalizingly close to in vitro-generated HSCs. Deciphering how to arrive there without transgenes but rather through the sequential addition of growth factors to guide the differentiating cells will be the ultimate demonstration that we understand the biology of making HSCs and advancements in this direction will bring us closer to fulfilling the enormous potential that iPSCs offer for regenerative medicine. Nevertheless, despite the current limitations of in vitro directed differentiation in generating bona fide HSCs, there exist many examples of how patient-specific iPSCs have been successfully used for the modeling of hematological disorders and how they have played a prominent role in uncovering pharmacologically targetable disease mechanisms.

References

1. Orkin SH, Zon LI. Hematopoiesis: an evolving paradigm for stem cell biology. Cell. 2008;132:631–44. https://doi.org/10.1016/j.cell.2008.01.025.
2. Vo LT, Daley GQ. De novo generation of HSCs from somatic and pluripotent stem cell sources. Blood. 2015;125:2641–8. https://doi.org/10.1182/blood-2014-10-570234.
3. Tavian M, Peault B. Embryonic development of the human hematopoietic system. Int J Dev Biol. 2005;49:243–50. https://doi.org/10.1387/ijdb.041957mt.
4. Ivanovs A, et al. Human haematopoietic stem cell development: from the embryo to the dish. Development. 2017;144:2323–37. https://doi.org/10.1242/dev.134866.
5. Ferkowicz MJ, Yoder MC. Blood island formation: longstanding observations and modern interpretations. Exp Hematol. 2005;33:1041–7. https://doi.org/10.1016/j.exphem.2005.06.006.
6. Palis J, Robertson S, Kennedy M, Wall C, Keller G. Development of erythroid and myeloid progenitors in the yolk sac and embryo proper of the mouse. Development. 1999;126:5073–84.
7. Wood WG. Haemoglobin synthesis during human fetal development. Br Med Bull. 1976;32:282–7.
8. Palis J. Primitive and definitive erythropoiesis in mammals. Front Physiol. 2014;5:3. https://doi.org/10.3389/fphys.2014.00003.
9. Philipsen S, Wood WG. In: Forget BG, Steinberg MH, Higgs DR, Weatherall DJ, editors. Disorders of hemoglobin: genetics, pathophysiology, and clinical management Ch. 2. Cambridge: Cambridge University Press; 2009. p. 24–45.
10. Fraser ST, Isern J, Baron MH. Maturation and enucleation of primitive erythroblasts during mouse embryogenesis is accompanied by changes in cell-surface antigen expression. Blood. 2007;109:343–52. https://doi.org/10.1182/blood-2006-03-006569.
11. Kingsley PD, Malik J, Fantauzzo KA, Palis J. Yolk sac-derived primitive erythroblasts enucleate during mammalian embryogenesis. Blood. 2004;104:19–25. https://doi.org/10.1182/blood-2003-12-4162.
12. Tober J, et al. The megakaryocyte lineage originates from hemangioblast precursors and is an integral component both of primitive and of definitive hematopoiesis. Blood. 2007;109:1433–41. https://doi.org/10.1182/blood-2006-06-031898.
13. Xu MJ, et al. Evidence for the presence of murine primitive megakaryocytopoiesis in the early yolk sac. Blood. 2001;97:2016–22.
14. Naito M, Yamamura F, Nishikawa S, Takahashi K. Development, differentiation, and maturation of fetal mouse yolk sac macrophages in cultures. J Leukoc Biol. 1989;46:1–10.
15. Takahashi K, Yamamura F, Naito M. Differentiation, maturation, and proliferation of macrophages in the mouse yolk sac: a light-microscopic, enzyme-cytochemical, immunohistochemical, and ultrastructural study. J Leukoc Biol. 1989;45:87–96.
16. Palis J, et al. Spatial and temporal emergence of high proliferative potential hematopoietic precursors during murine embryogenesis. Proc Natl Acad Sci U S A. 2001;98:4528–33. https://doi.org/10.1073/pnas.071002398.

17. Ditadi A, Sturgeon CM, Keller G. A view of human haematopoietic development from the Petri dish. Nat Rev Mol Cell Biol. 2017;18:56–67. https://doi.org/10.1038/nrm.2016.127.
18. McGrath KE, et al. A transient definitive erythroid lineage with unique regulation of the beta-globin locus in the mammalian embryo. Blood. 2011;117:4600–8. https://doi.org/10.1182/blood-2010-12-325357.
19. Migliaccio G, et al. Human embryonic hemopoiesis. Kinetics of progenitors and precursors underlying the yolk sac----liver transition. J Clin Invest. 1986;78:51–60. https://doi.org/10.1172/JCI112572.
20. Yoshimoto M, et al. Embryonic day 9 yolk sac and intra-embryonic hemogenic endothelium independently generate a B-1 and marginal zone progenitor lacking B-2 potential. Proc Natl Acad Sci U S A. 2011;108:1468–73. https://doi.org/10.1073/pnas.1015841108.
21. Yoshimoto M, et al. Autonomous murine T-cell progenitor production in the extra-embryonic yolk sac before HSC emergence. Blood. 2012;119:5706–14. https://doi.org/10.1182/blood-2011-12-397489.
22. Boiers C, et al. Lymphomyeloid contribution of an immune-restricted progenitor emerging prior to definitive hematopoietic stem cells. Cell Stem Cell. 2013;13:535–48. https://doi.org/10.1016/j.stem.2013.08.012.
23. Montecino-Rodriguez E, Dorshkind K. B-1 B cell development in the fetus and adult. Immunity. 2012;36:13–21. https://doi.org/10.1016/j.immuni.2011.11.017.
24. Ginhoux F, Guilliams M. Tissue-resident macrophage ontogeny and homeostasis. Immunity. 2016;44:439–49. https://doi.org/10.1016/j.immuni.2016.02.024.
25. Gomez Perdiguero E, et al. Tissue-resident macrophages originate from yolk-sac-derived erythro-myeloid progenitors. Nature. 2015;518:547–51. https://doi.org/10.1038/nature13989.
26. Hoeffel G, et al. C-Myb(+) erythro-myeloid progenitor-derived fetal monocytes give rise to adult tissue-resident macrophages. Immunity. 2015;42:665–78. https://doi.org/10.1016/j.immuni.2015.03.011.
27. Schulz C, et al. A lineage of myeloid cells independent of Myb and hematopoietic stem cells. Science. 2012;336:86–90. https://doi.org/10.1126/science.1219179.
28. Muller AM, Medvinsky A, Strouboulis J, Grosveld F, Dzierzak E. Development of hematopoietic stem cell activity in the mouse embryo. Immunity. 1994;1:291–301.
29. Medvinsky A, Dzierzak E. Definitive hematopoiesis is autonomously initiated by the AGM region. Cell. 1996;86:897–906.
30. Ivanovs A, et al. Highly potent human hematopoietic stem cells first emerge in the intra-embryonic aorta-gonad-mesonephros region. J Exp Med. 2011;208:2417–27. https://doi.org/10.1084/jem.20111688.
31. de Bruijn MF, et al. Hematopoietic stem cells localize to the endothelial cell layer in the midgestation mouse aorta. Immunity. 2002;16:673–83.
32. de Bruijn MF, Speck NA, Peeters MC, Dzierzak E. Definitive hematopoietic stem cells first develop within the major arterial regions of the mouse embryo. EMBO J. 2000;19:2465–74. https://doi.org/10.1093/emboj/19.11.2465.
33. Gekas C, Dieterlen-Lievre F, Orkin SH, Mikkola HK. The placenta is a niche for hematopoietic stem cells. Dev Cell. 2005;8:365–75. https://doi.org/10.1016/j.devcel.2004.12.016.
34. Ottersbach K, Dzierzak E. The murine placenta contains hematopoietic stem cells within the vascular labyrinth region. Dev Cell. 2005;8:377–87. https://doi.org/10.1016/j.devcel.2005.02.001.
35. Medvinsky A, Rybtsov S, Taoudi S. Embryonic origin of the adult hematopoietic system: advances and questions. Development. 2011;138:1017–31. https://doi.org/10.1242/dev.040998.
36. Charbord P, Tavian M, Humeau L, Peault B. Early ontogeny of the human marrow from long bones: an immunohistochemical study of hematopoiesis and its microenvironment. Blood. 1996;87:4109–19.
37. Blazsek I, Chagraoui J, Peault B. Ontogenic emergence of the hematon, a morphogenetic stromal unit that supports multipotential hematopoietic progenitors in mouse bone marrow. Blood. 2000;96:3763–71.

38. Dzierzak E, Speck NA. Of lineage and legacy: the development of mammalian hematopoietic stem cells. Nat Immunol. 2008;9:129–36. https://doi.org/10.1038/ni1560.
39. North TE, et al. Runx1 expression marks long-term repopulating hematopoietic stem cells in the midgestation mouse embryo. Immunity. 2002;16:661–72.
40. Taoudi S, Medvinsky A. Functional identification of the hematopoietic stem cell niche in the ventral domain of the embryonic dorsal aorta. Proc Natl Acad Sci U S A. 2007;104:9399–403. https://doi.org/10.1073/pnas.0700984104.
41. Boisset JC, et al. In vivo imaging of haematopoietic cells emerging from the mouse aortic endothelium. Nature. 2010;464:116–20. https://doi.org/10.1038/nature08764.
42. Bertrand JY, et al. Haematopoietic stem cells derive directly from aortic endothelium during development. Nature. 2010;464:108–11. https://doi.org/10.1038/nature08738.
43. Kissa K, Herbomel P. Blood stem cells emerge from aortic endothelium by a novel type of cell transition. Nature. 2010;464:112–5. https://doi.org/10.1038/nature08761.
44. Frame JM, Fegan KH, Conway SJ, McGrath KE, Palis J. Definitive hematopoiesis in the yolk sac emerges from Wnt-responsive hemogenic endothelium independently of circulation and arterial identity. Stem Cells. 2016;34:431–44. https://doi.org/10.1002/stem.2213.
45. Samokhvalov IM, Samokhvalova NI, Nishikawa S. Cell tracing shows the contribution of the yolk sac to adult haematopoiesis. Nature. 2007;446:1056–61. https://doi.org/10.1038/nature05725.
46. Tanaka Y, et al. Early ontogenic origin of the hematopoietic stem cell lineage. Proc Natl Acad Sci U S A. 2012;109:4515–20. https://doi.org/10.1073/pnas.1115828109.
47. Tanaka Y, et al. Circulation-independent differentiation pathway from extraembryonic mesoderm toward hematopoietic stem cells via hemogenic angioblasts. Cell Rep. 2014;8:31–9. https://doi.org/10.1016/j.celrep.2014.05.055.
48. Eliades A, et al. The hemogenic competence of endothelial progenitors is restricted by Runx1 silencing during embryonic development. Cell Rep. 2016;15:2185–99. https://doi.org/10.1016/j.celrep.2016.05.001.
49. Jaffredo T, Gautier R, Eichmann A, Dieterlen-Lievre F. Intraaortic hemopoietic cells are derived from endothelial cells during ontogeny. Development. 1998;125:4575–83.
50. Ciau-Uitz A, Walmsley M, Patient R. Distinct origins of adult and embryonic blood in Xenopus. Cell. 2000;102:787–96.
51. Dieterlen-Lievre F, Martin C. Diffuse intraembryonic hemopoiesis in normal and chimeric avian development. Dev Biol. 1981;88:180–91.
52. Garcia-Porrero JA, Godin IE, Dieterlen-Lievre F. Potential intraembryonic hemogenic sites at pre-liver stages in the mouse. Anat Embryol (Berl). 1995;192:425–35.
53. Pardanaud L, Yassine F, Dieterlen-Lievre F. Relationship between vasculogenesis, angiogenesis and haemopoiesis during avian ontogeny. Development. 1989;105:473–85.
54. Tavian M, et al. Aorta-associated CD34+ hematopoietic cells in the early human embryo. Blood. 1996;87:67–72.
55. Wood HB, May G, Healy L, Enver T, Morriss-Kay GM. CD34 expression patterns during early mouse development are related to modes of blood vessel formation and reveal additional sites of hematopoiesis. Blood. 1997;90:2300–11.
56. Yoder MC, et al. Characterization of definitive lymphohematopoietic stem cells in the day 9 murine yolk sac. Immunity. 1997;7:335–44.
57. Yoder MC, Hiatt K, Mukherjee P. In vivo repopulating hematopoietic stem cells are present in the murine yolk sac at day 9.0 postcoitus. Proc Natl Acad Sci U S A. 1997;94:6776–80.
58. Taoudi S, et al. Extensive hematopoietic stem cell generation in the AGM region via maturation of VE-cadherin+CD45+ pre-definitive HSCs. Cell Stem Cell. 2008;3:99–108. https://doi.org/10.1016/j.stem.2008.06.004.
59. Rybtsov S, et al. Tracing the origin of the HSC hierarchy reveals an SCF-dependent, IL-3-independent CD43(−) embryonic precursor. Stem Cell Rep. 2014;3:489–501. https://doi.org/10.1016/j.stemcr.2014.07.009.

60. Rybtsov S, et al. Hierarchical organization and early hematopoietic specification of the developing HSC lineage in the AGM region. J Exp Med. 2011;208:1305–15. https://doi.org/10.1084/jem.20102419.
61. Dieterlen-Lievre F. On the origin of haemopoietic stem cells in the avian embryo: an experimental approach. J Embryol Exp Morphol. 1975;33:607–19.
62. Lux CT, et al. All primitive and definitive hematopoietic progenitor cells emerging before E10 in the mouse embryo are products of the yolk sac. Blood. 2008;111:3435–8. https://doi.org/10.1182/blood-2007-08-107086.
63. Gritz E, Hirschi KK. Specification and function of hemogenic endothelium during embryogenesis. Cell Mol Life Sci. 2016;73:1547–67. https://doi.org/10.1007/s00018-016-2134-0.
64. Sturgeon CM, Ditadi A, Awong G, Kennedy M, Keller G. Wnt signaling controls the specification of definitive and primitive hematopoiesis from human pluripotent stem cells. Nat Biotechnol. 2014;32:554–61. https://doi.org/10.1038/nbt.2915.
65. Lacaud G, Kouskoff V. Hemangioblast, hemogenic endothelium, and primitive versus definitive hematopoiesis. Exp Hematol. 2017;49:19–24. https://doi.org/10.1016/j.exphem.2016.12.009.
66. Kennedy M, et al. T lymphocyte potential marks the emergence of definitive hematopoietic progenitors in human pluripotent stem cell differentiation cultures. Cell Rep. 2012;2:1722–35. https://doi.org/10.1016/j.celrep.2012.11.003.
67. Tian Y, et al. The first wave of T lymphopoiesis in zebrafish arises from aorta endothelium independent of hematopoietic stem cells. J Exp Med. 2017;214(11):3347. https://doi.org/10.1084/jem.20170488.
68. Dzierzak E, Philipsen S. Erythropoiesis: development and differentiation. Cold Spring Harb Perspect Med. 2013;3:a011601. https://doi.org/10.1101/cshperspect.a011601.
69. Zambidis ET, Peault B, Park TS, Bunz F, Civin CI. Hematopoietic differentiation of human embryonic stem cells progresses through sequential hematoendothelial, primitive, and definitive stages resembling human yolk sac development. Blood. 2005;106:860–70. https://doi.org/10.1182/blood-2004-11-4522.
70. Murry CE, Keller G. Differentiation of embryonic stem cells to clinically relevant populations: lessons from embryonic development. Cell. 2008;132:661–80. https://doi.org/10.1016/j.cell.2008.02.008.
71. Kaufman DS, Hanson ET, Lewis RL, Auerbach R, Thomson JA. Hematopoietic colony-forming cells derived from human embryonic stem cells. Proc Natl Acad Sci U S A. 2001;98:10716–21. https://doi.org/10.1073/pnas.191362598.
72. Vodyanik MA, Bork JA, Thomson JA, Slukvin II. Human embryonic stem cell-derived CD34+ cells: efficient production in the coculture with OP9 stromal cells and analysis of lymphohematopoietic potential. Blood. 2005;105:617–26. https://doi.org/10.1182/blood-2004-04-1649.
73. Vodyanik MA, Thomson JA, Slukvin II. Leukosialin (CD43) defines hematopoietic progenitors in human embryonic stem cell differentiation cultures. Blood. 2006;108:2095–105. https://doi.org/10.1182/blood-2006-02-003327.
74. Woll PS, et al. Wnt signaling promotes hematoendothelial cell development from human embryonic stem cells. Blood. 2008;111:122–31. https://doi.org/10.1182/blood-2007-04-084186.
75. Ji J, et al. OP9 stroma augments survival of hematopoietic precursors and progenitors during hematopoietic differentiation from human embryonic stem cells. Stem Cells. 2008;26:2485–95. https://doi.org/10.1634/stemcells.2008-0642.
76. Ledran MH, et al. Efficient hematopoietic differentiation of human embryonic stem cells on stromal cells derived from hematopoietic niches. Cell Stem Cell. 2008;3:85–98. https://doi.org/10.1016/j.stem.2008.06.001.
77. Qiu C, Olivier EN, Velho M, Bouhassira EE. Globin switches in yolk sac-like primitive and fetal-like definitive red blood cells produced from human embryonic stem cells. Blood. 2008;111:2400–8. https://doi.org/10.1182/blood-2007-07-102087.
78. Ochi K, et al. Multicolor staining of globin subtypes reveals impaired globin switching during erythropoiesis in human pluripotent stem cells. Stem Cells Transl Med. 2014;3:792–800. https://doi.org/10.5966/sctm.2013-0216.

79. Chadwick K, et al. Cytokines and BMP-4 promote hematopoietic differentiation of human embryonic stem cells. Blood. 2003;102:906–15. https://doi.org/10.1182/blood-2003-03-0832.
80. Ng ES, Davis RP, Azzola L, Stanley EG, Elefanty AG. Forced aggregation of defined numbers of human embryonic stem cells into embryoid bodies fosters robust, reproducible hematopoietic differentiation. Blood. 2005;106:1601–3. https://doi.org/10.1182/blood-2005-03-0987.
81. Kennedy M, D'Souza SL, Lynch-Kattman M, Schwantz S, Keller G. Development of the hemangioblast defines the onset of hematopoiesis in human ES cell differentiation cultures. Blood. 2007;109:2679–87. https://doi.org/10.1182/blood-2006-09-047704.
82. Niwa A, et al. A novel serum-free monolayer culture for orderly hematopoietic differentiation of human pluripotent cells via mesodermal progenitors. PLoS One. 2011;6:e22261. https://doi.org/10.1371/journal.pone.0022261.
83. Smith BW, et al. The aryl hydrocarbon receptor directs hematopoietic progenitor cell expansion and differentiation. Blood. 2013;122:376–85. https://doi.org/10.1182/blood-2012-11-466722.
84. Leung A, et al. Notch and aryl hydrocarbon receptor signaling impact definitive hematopoiesis from human pluripotent stem cells. Stem Cells. 2018;36(7):1004–1019. https://doi.org/10.1002/stem.2822. Epub 2018 Apr 1.
85. Abazov VM, et al. Ratio of isolated photon cross sections in pp macro collisions at square root of s = 630 and 1800 GeV. Phys Rev Lett. 2001;87:251805.
86. Davis RP, et al. Targeting a GFP reporter gene to the MIXL1 locus of human embryonic stem cells identifies human primitive streak-like cells and enables isolation of primitive hematopoietic precursors. Blood. 2008;111:1876–84. https://doi.org/10.1182/blood-2007-06-093609.
87. Pick M, Azzola L, Mossman A, Stanley EG, Elefanty AG. Differentiation of human embryonic stem cells in serum-free medium reveals distinct roles for bone morphogenetic protein 4, vascular endothelial growth factor, stem cell factor, and fibroblast growth factor 2 in hematopoiesis. Stem Cells. 2007;25:2206–14. https://doi.org/10.1634/stemcells.2006-0713.
88. Woll PS, Martin CH, Miller JS, Kaufman DS. Human embryonic stem cell-derived NK cells acquire functional receptors and cytolytic activity. J Immunol. 2005;175:5095–103.
89. Jung HS, et al. A human VE-cadherin-tdTomato and CD43-green fluorescent protein dual reporter cell line for study endothelial to hematopoietic transition. Stem Cell Res. 2016;17:401–5. https://doi.org/10.1016/j.scr.2016.09.004.
90. Wang Y, Nakayama N. WNT and BMP signaling are both required for hematopoietic cell development from human ES cells. Stem Cell Res. 2009;3:113–25. https://doi.org/10.1016/j.scr.2009.06.001.
91. Bunn HF. Erythropoietin. Cold Spring Harb Perspect Med. 2013;3:a011619. https://doi.org/10.1101/cshperspect.a011619.
92. Radtke F, et al. Deficient T cell fate specification in mice with an induced inactivation of Notch1. Immunity. 1999;10:547–58.
93. Allman D, Aster JC, Pear WS. Notch signaling in hematopoiesis and early lymphocyte development. Immunol Rev. 2002;187:75–86.
94. Deutsch VR, Tomer A. Megakaryocyte development and platelet production. Br J Haematol. 2006;134:453–66. https://doi.org/10.1111/j.1365-2141.2006.06215.x.
95. Yoshihara H, et al. Thrombopoietin/MPL signaling regulates hematopoietic stem cell quiescence and interaction with the osteoblastic niche. Cell Stem Cell. 2007;1:685–97. https://doi.org/10.1016/j.stem.2007.10.020.
96. Irion S, et al. Temporal specification of blood progenitors from mouse embryonic stem cells and induced pluripotent stem cells. Development. 2010;137:2829–39. https://doi.org/10.1242/dev.042119.
97. Choi KD, et al. Identification of the hemogenic endothelial progenitor and its direct precursor in human pluripotent stem cell differentiation cultures. Cell Rep. 2012;2:553–67. https://doi.org/10.1016/j.celrep.2012.08.002.
98. Zambidis ET, et al. Expression of angiotensin-converting enzyme (CD143) identifies and regulates primitive hemangioblasts derived from human pluripotent stem cells. Blood. 2008;112:3601–14. https://doi.org/10.1182/blood-2008-03-144766.

99. Ng ES, et al. Differentiation of human embryonic stem cells to HOXA+ hemogenic vasculature that resembles the aorta-gonad-mesonephros. Nat Biotechnol. 2016;34:1168–79. https://doi.org/10.1038/nbt.3702.
100. Lu SJ, et al. Generation of functional hemangioblasts from human embryonic stem cells. Nat Methods. 2007;4:501–9. https://doi.org/10.1038/nmeth1041.
101. Choi K. Hemangioblast development and regulation. Biochem Cell Biol. 1998;76:947–56.
102. Fehling HJ, et al. Tracking mesoderm induction and its specification to the hemangioblast during embryonic stem cell differentiation. Development. 2003;130:4217–27.
103. Huber TL, Kouskoff V, Fehling HJ, Palis J, Keller G. Haemangioblast commitment is initiated in the primitive streak of the mouse embryo. Nature. 2004;432:625–30. https://doi.org/10.1038/nature03122.
104. North T, et al. Cbfa2 is required for the formation of intra-aortic hematopoietic clusters. Development. 1999;126:2563–75.
105. Thambyrajah R, et al. New insights into the regulation by RUNX1 and GFI1(s) proteins of the endothelial to hematopoietic transition generating primordial hematopoietic cells. Cell Cycle. 2016;15:2108–14. https://doi.org/10.1080/15384101.2016.1203491.
106. Ditadi A, et al. Human definitive haemogenic endothelium and arterial vascular endothelium represent distinct lineages. Nat Cell Biol. 2015;17:580–91. https://doi.org/10.1038/ncb3161.
107. Notta F, et al. Isolation of single human hematopoietic stem cells capable of long-term multilineage engraftment. Science. 2011;333:218–21. https://doi.org/10.1126/science.1201219.
108. Shultz LD, Brehm MA, Garcia-Martinez JV, Greiner DL. Humanized mice for immune system investigation: progress, promise and challenges. Nat Rev Immunol. 2012;12:786–98. https://doi.org/10.1038/nri3311.
109. Ito M, et al. NOD/SCID/gamma(c)(null) mouse: an excellent recipient mouse model for engraftment of human cells. Blood. 2002;100:3175–82. https://doi.org/10.1182/blood-2001-12-0207.
110. McDermott SP, Eppert K, Lechman ER, Doedens M, Dick JE. Comparison of human cord blood engraftment between immunocompromised mouse strains. Blood. 2010;116:193–200. https://doi.org/10.1182/blood-2010-02-271841.
111. Lu M, Kardel MD, O'Connor MD, Eaves CJ. Enhanced generation of hematopoietic cells from human hepatocarcinoma cell-stimulated human embryonic and induced pluripotent stem cells. Exp Hematol. 2009;37:924–36. https://doi.org/10.1016/j.exphem.2009.05.007.
112. Wang L, et al. Generation of hematopoietic repopulating cells from human embryonic stem cells independent of ectopic HOXB4 expression. J Exp Med. 2005;201:1603–14. https://doi.org/10.1084/jem.20041888.
113. Suzuki N, et al. Generation of engraftable hematopoietic stem cells from induced pluripotent stem cells by way of teratoma formation. Mol Ther. 2013;21:1424–31. https://doi.org/10.1038/mt.2013.71.
114. Amabile G, et al. In vivo generation of transplantable human hematopoietic cells from induced pluripotent stem cells. Blood. 2013;121:1255–64. https://doi.org/10.1182/blood-2012-06-434407.
115. Hentze H, et al. Teratoma formation by human embryonic stem cells: evaluation of essential parameters for future safety studies. Stem Cell Res. 2009;2:198–210. https://doi.org/10.1016/j.scr.2009.02.002.
116. Tsukada M, et al. In Vivo Generation of Engraftable Murine Hematopoietic Stem Cells by Gfi1b, c-Fos, and Gata2 Overexpression within Teratoma. Stem Cell Rep. 2017;9:1024–33. https://doi.org/10.1016/j.stemcr.2017.08.010.
117. Yamanaka S, Takahashi K. Induction of pluripotent stem cells from mouse fibroblast cultures. Tanpakushitsu Kakusan Koso. 2006;51:2346–51.
118. Kyba M, Perlingeiro RC, Daley GQ. HoxB4 confers definitive lymphoid-myeloid engraftment potential on embryonic stem cell and yolk sac hematopoietic progenitors. Cell. 2002;109:29–37.

119. Elcheva I, et al. Direct induction of haematoendothelial programs in human pluripotent stem cells by transcriptional regulators. Nat Commun. 2014;5:4372. https://doi.org/10.1038/ncomms5372.
120. Pereira CF, et al. Induction of a hemogenic program in mouse fibroblasts. Cell Stem Cell. 2013;13:205–18. https://doi.org/10.1016/j.stem.2013.05.024.
121. Batta K, Florkowska M, Kouskoff V, Lacaud G. Direct reprogramming of murine fibroblasts to hematopoietic progenitor cells. Cell Rep. 2014;9:1871–84. https://doi.org/10.1016/j.celrep.2014.11.002.
122. Doulatov S, et al. Induction of multipotential hematopoietic progenitors from human pluripotent stem cells via respecification of lineage-restricted precursors. Cell Stem Cell. 2013;13:459–70. https://doi.org/10.1016/j.stem.2013.09.002.
123. Riddell J, et al. Reprogramming committed murine blood cells to induced hematopoietic stem cells with defined factors. Cell. 2014;157:549–64. https://doi.org/10.1016/j.cell.2014.04.006.
124. Sandler VM, et al. Reprogramming human endothelial cells to haematopoietic cells requires vascular induction. Nature. 2014;511:312–8. https://doi.org/10.1038/nature13547.
125. Lis R, et al. Conversion of adult endothelium to immunocompetent haematopoietic stem cells. Nature. 2017;545:439–45. https://doi.org/10.1038/nature22326.
126. Sugimura R, et al. Haematopoietic stem and progenitor cells from human pluripotent stem cells. Nature. 2017;545:432–8. https://doi.org/10.1038/nature22370.
127. Stamatoyannopoulos G. Control of globin gene expression during development and erythroid differentiation. Exp Hematol. 2005;33:259–71. https://doi.org/10.1016/j.exphem.2004.11.007.
128. Bauer DE, Kamran SC, Orkin SH. Reawakening fetal hemoglobin: prospects for new therapies for the beta-globin disorders. Blood. 2012;120:2945–53. https://doi.org/10.1182/blood-2012-06-292078.
129. Sankaran VG, Xu J, Orkin SH. Advances in the understanding of haemoglobin switching. Br J Haematol. 2010;149:181–94. https://doi.org/10.1111/j.1365-2141.2010.08105.x.
130. Peschle C, et al. Haemoglobin switching in human embryos: asynchrony of zeta----alpha and epsilon----gamma-globin switches in primitive and definite erythropoietic lineage. Nature. 1985;313:235–8.
131. Yang CT, et al. Human induced pluripotent stem cell derived erythroblasts can undergo definitive erythropoiesis and co-express gamma and beta globins. Br J Haematol. 2014;166:435–48. https://doi.org/10.1111/bjh.12910.
132. Patterson M, et al. Defining the nature of human pluripotent stem cell progeny. Cell Res. 2012;22:178–93. https://doi.org/10.1038/cr.2011.133.
133. Vanuytsel K, et al. Induced pluripotent stem cell-based mapping of β -globin expression throughout human erythropoietic development. Blood Adv. 2018;2(15):1998–2011. https://doi.org/10.1182/bloodadvances.2018020560
134. Kelley JM, Daley GQ. Hematopoietic defects and iPSC disease modeling: lessons learned. Immunol Lett. 2013;155:18–20. https://doi.org/10.1016/j.imlet.2013.09.018.
135. Ceccaldi R, et al. Bone marrow failure in Fanconi anemia is triggered by an exacerbated p53/p21 DNA damage response that impairs hematopoietic stem and progenitor cells. Cell Stem Cell. 2012;11:36–49. https://doi.org/10.1016/j.stem.2012.05.013.
136. Tulpule A, et al. Knockdown of Fanconi anemia genes in human embryonic stem cells reveals early developmental defects in the hematopoietic lineage. Blood. 2010;115:3453–62. https://doi.org/10.1182/blood-2009-10-246694.
137. Tanno T, et al. High levels of GDF15 in thalassemia suppress expression of the iron regulatory protein hepcidin. Nat Med. 2007;13:1096–101. https://doi.org/10.1038/nm1629.
138. Tanno T, et al. Identification of TWSG1 as a second novel erythroid regulator of hepcidin expression in murine and human cells. Blood. 2009;114:181–6. https://doi.org/10.1182/blood-2008-12-195503.

139. Kautz L, et al. Identification of erythroferrone as an erythroid regulator of iron metabolism. Nat Genet. 2014;46:678–84. https://doi.org/10.1038/ng.2996.
140. Agarwal S, et al. Telomere elongation in induced pluripotent stem cells from dyskeratosis congenita patients. Nature. 2010;464:292–6. https://doi.org/10.1038/nature08792.
141. Muller LU, et al. Overcoming reprogramming resistance of Fanconi anemia cells. Blood. 2012;119:5449–57. https://doi.org/10.1182/blood-2012-02-408674.
142. Tulpule A, et al. Pluripotent stem cell models of Shwachman-Diamond syndrome reveal a common mechanism for pancreatic and hematopoietic dysfunction. Cell Stem Cell. 2013;12:727–36. https://doi.org/10.1016/j.stem.2013.04.002.
143. Cherry AB, et al. Induced pluripotent stem cells with a mitochondrial DNA deletion. Stem Cells. 2013;31:1287–97. https://doi.org/10.1002/stem.1354.
144. Garcon L, et al. Ribosomal and hematopoietic defects in induced pluripotent stem cells derived from Diamond Blackfan anemia patients. Blood. 2013;122:912–21. https://doi.org/10.1182/blood-2013-01-478321.
145. Zou J, Mali P, Huang X, Dowey SN, Cheng L. Site-specific gene correction of a point mutation in human iPS cells derived from an adult patient with sickle cell disease. Blood. 2011;118:4599–608. https://doi.org/10.1182/blood-2011-02-335554.
146. Park S, et al. A Comprehensive, ethnically diverse library of sickle cell disease-specific induced pluripotent stem cells. Stem Cell Rep. 2017;8:1076–85. https://doi.org/10.1016/j.stemcr.2016.12.017.
147. Li C, et al. Novel HDAd/EBV reprogramming vector and highly efficient Ad/CRISPR-cas sickle cell disease gene correction. Sci Rep. 2016;6:30422. https://doi.org/10.1038/srep30422.
148. Xie F, et al. Seamless gene correction of beta-thalassemia mutations in patient-specific iPSCs using CRISPR/Cas9 and piggyBac. Genome Res. 2014;24:1526–33. https://doi.org/10.1101/gr.173427.114.
149. Ou Z, et al. The combination of CRISPR/Cas9 and iPSC technologies in the gene therapy of human beta-thalassemia in mice. Sci Rep. 2016;6:32463. https://doi.org/10.1038/srep32463.
150. Niu X, et al. Combining single strand oligodeoxynucleotides and CRISPR/Cas9 to correct gene mutations in beta-thalassemia-induced pluripotent stem cells. J Biol Chem. 2016;291:16576–85. https://doi.org/10.1074/jbc.M116.719237.
151. Phanthong P, et al. Enhancement of beta-globin gene expression in thalassemic IVS2-654 induced pluripotent stem cell-derived erythroid cells by modified U7 snRNA. Stem Cells Transl Med. 2017;6:1059–69. https://doi.org/10.1002/sctm.16-0121.
152. Kumano K, et al. Generation of induced pluripotent stem cells from primary chronic myelogenous leukemia patient samples. Blood. 2012;119:6234–42. https://doi.org/10.1182/blood-2011-07-367441.
153. Fok WC, et al. p53 mediates failure of human definitive hematopoiesis in dyskeratosis congenita. Stem Cell Rep. 2017;9:409–18. https://doi.org/10.1016/j.stemcr.2017.06.015.
154. Raya A, et al. Disease-corrected haematopoietic progenitors from Fanconi anaemia induced pluripotent stem cells. Nature. 2009;460:53–9. https://doi.org/10.1038/nature08129.
155. Chlon TM, et al. Overcoming pluripotent stem cell dependence on the repair of endogenous DNA damage. Stem Cell Rep. 2016;6:44–54. https://doi.org/10.1016/j.stemcr.2015.12.001.
156. Yung SK, et al. Brief report: human pluripotent stem cell models of fanconi anemia deficiency reveal an important role for fanconi anemia proteins in cellular reprogramming and survival of hematopoietic progenitors. Stem Cells. 2013;31:1022–9. https://doi.org/10.1002/stem.1308.
157. Vanuytsel K, et al. FANCA knockout in human embryonic stem cells causes a severe growth disadvantage. Stem Cell Res. 2014;13:240–50. https://doi.org/10.1016/j.scr.2014.07.005.
158. Liu GH, et al. Modelling Fanconi anemia pathogenesis and therapeutics using integration-free patient-derived iPSCs. Nat Commun. 2014;5:4330. https://doi.org/10.1038/ncomms5330.
159. Sankaran VG, Orkin SH. The switch from fetal to adult hemoglobin. Cold Spring Harb Perspect Med. 2013;3:a011643. https://doi.org/10.1101/cshperspect.a011643.

160. Steinberg MH, Ohene-Frempong K, Heeney MM. In: Forget BG, Steinberg MH, Higgs DR, Weatherall DJ, editors. Disorders of hemoglobin: genetics, pathophysiology, and clinical management Ch. 19. Cambridge: Cambridge University Press; 2009. p. 437–496.
161. Chang KH, et al. Definitive-like erythroid cells derived from human embryonic stem cells coexpress high levels of embryonic and fetal globins with little or no adult globin. Blood. 2006;108:1515–23. https://doi.org/10.1182/blood-2005-11-011874.
162. Platt OS, et al. Mortality in sickle cell disease. Life expectancy and risk factors for early death. N Engl J Med. 1994;330:1639–44. https://doi.org/10.1056/NEJM199406093302303.
163. Platt OS, et al. Pain in sickle cell disease. Rates and risk factors. N Engl J Med. 1991;325:11–6. https://doi.org/10.1056/NEJM199107043250103.
164. Piel FB, Steinberg MH, Rees DC. Sickle Cell Disease. N Engl J Med. 2017;377:305. https://doi.org/10.1056/NEJMc1706325.
165. Stamatoyannopoulos G, Navas PA, Li Q. In: Forget BG, Steinberg MH, Higgs DR, Weatherall DJ, editors. Disorders of hemoglobin: genetics, pathophysiology, and clinical management Ch. 5. Cambridge: Cambridge University Press; 2009. p. 86–100.
166. Rahmig S, et al. Improved human erythropoiesis and platelet formation in humanized NSGW41 mice. Stem Cell Rep. 2016;7:591–601. https://doi.org/10.1016/j.stemcr.2016.08.005.
167. Chen B, et al. Complement depletion improves human red blood cell reconstitution in immunodeficient mice. Stem Cell Rep. 2017;9:1034–42. https://doi.org/10.1016/j.stemcr.2017.08.018.
168. Fiorini C, et al. Developmentally-faithful and effective human erythropoiesis in immunodeficient and Kit mutant mice. Am J Hematol. 2017;92:E513–9. https://doi.org/10.1002/ajh.24805.
169. Hu Z, Van Rooijen N, Yang YG. Macrophages prevent human red blood cell reconstitution in immunodeficient mice. Blood. 2011;118:5938–46. https://doi.org/10.1182/blood-2010-11-321414.
170. McIntosh BE, et al. Nonirradiated NOD,B6.SCID Il2rgamma−/− Kit(W41/W41) (NBSGW) mice support multilineage engraftment of human hematopoietic cells. Stem Cell Rep. 2015;4:171–80. https://doi.org/10.1016/j.stemcr.2014.12.005.
171. Steinberg MH, Chui DH, Dover GJ, Sebastiani P, Alsultan A. Fetal hemoglobin in sickle cell anemia: a glass half full? Blood. 2014;123:481–5. https://doi.org/10.1182/blood-2013-09-528067.
172. Kola I, Landis J. Can the pharmaceutical industry reduce attrition rates? Nat Rev Drug Discov. 2004;3:711–5. https://doi.org/10.1038/nrd1470.
173. Lian Q, Chow Y, Esteban MA, Pei D, Tse HF. Future perspective of induced pluripotent stem cells for diagnosis, drug screening and treatment of human diseases. Thromb Haemost. 2010;104:39–44. https://doi.org/10.1160/TH10-05-0269.
174. Deshmukh RS, Kovacs KA, Dinnyes A. Drug discovery models and toxicity testing using embryonic and induced pluripotent stem cell-derived cardiac and neuronal cells. Stem Cells Int. 2012;2012:379569. https://doi.org/10.1155/2012/379569.
175. Chun YS, Byun K, Lee B. Induced pluripotent stem cells and personalized medicine: current progress and future perspectives. Anat Cell Biol. 2011;44:245–55. https://doi.org/10.5115/acb.2011.44.4.245.

Chapter 5
Modeling Cardiomyopathies with iPSCs

Jean-Sébastien Hulot

Abstract Cardiomyopathies are disorders with primary defect of the cardiac muscle, typically presenting in a familial context and associated with mutations in major determinants of the electrical or contractile properties of cardiomyocytes. This chapter reviews the main advances in the physiopathological and pharmacological understanding of cardiomyopathies using iPSCs.

Keywords Cardiomyopathy · Heart failure · Arrhythmias · Sudden death Electrophysiology · Contractility

5.1 Introduction

Cardiomyopathy is a term used to define an injury of the cardiac muscle that can have multiple origins including extrinsic factors (such as ischemia in the context of a myocardial infarction or drug toxicities) or intrinsic factors (genetic mutations of key components of the heart). In clinical practice, "cardiomyopathies" however typically refer to primary cardiac muscle disorders, potentially presenting in a familial context and with particular morphological and/or physiological characteristics. The classification system of cardiomyopathies was originally conceived before the description of causative genetic defects and was essentially based on anatomical and histological observations [1]. The advances in molecular genetics, the progressive understanding of pathophysiological mechanisms, and the opportunity to model these diseases using iPSC technology are a source of perpetual evolution of the classification of cardiomyopathies [2]. Table 5.1 proposes a new classification based on historical and updated classifications as well as current understanding of primary molecular defect causing the cardiomyopathy. This classification will be further used in this chapter.

J.-S. Hulot (✉)
Paris Cardiovascular Research Center PARCC, INSERM UMR970, Hôpital Européen Georges Pompidou, Paris, France
e-mail: jean-sebastien.hulot@aphp.fr

H. Inoue, Y. Nakamura (eds.), *Medical Applications of iPS Cells*, Current Human Cell Research and Applications,
https://doi.org/10.1007/978-981-13-3672-0_5

Table 5.1 Proposed classification of cardiomyopathies

Causal primary defect	Principal outcome	Type of cardiomyopathy	Description
Cardiac ion channels	Life-threatening cardiac arrhythmias	Long QT syndrome	Delayed repolarization of the heart; Abnormal EKG
		Catecholaminergic polymorphic ventricular tachycardia	Stress-induced ventricular tachycardia
		Brugada syndrome	Abnormal EKG
Cardiomyocyte contractile machinery	Decreased contractility with heart failure; cardiac arrhythmias	Dilated cardiomyopathy	Dilation of cardiac cavities; shortness of breath, congestion
		Hypertrophic cardiomyopathy	Increase thickness of cardiac wall; abnormal EKG, familial cardiac sudden death
Cardiomyocyte architecture	Heart failure	Duchenne cardiomyopathy	Cardiac failure in the context of muscular dystrophy
Cardiomyocyte desmosome	Cardiac arrhythmias; right ventricular (RV) failure	Arrhythmogenic right ventricular cardiomyopathy	Arrhythmias and risk of sudden death, abnormal EKG; RV failure
Infiltrative disorder	Heart failure	Lysosomal storage disease (Fabry disease)	Heart failure with extra-cardiac manifestations

Historically, animal models have been used to understand the pathophysiology of cardiovascular diseases and discover new therapeutics; however, because of significant differences in cardiovascular genetics and physiology between humans and animals, there was a compelling need for more accurate models to understand cardiomyopathies. Over the past several years, the increasingly refined capacity to differentiate iPSCs into disease-relevant cell types, such as cardiomyocytes (iPSC-CMs), has offered new opportunities to model cardiomyopathies, to test effects of drugs or to develop cell-based therapies [3–6]. Since its first description in human embryonic stem cells [7, 8], cardiac differentiation of pluripotent stem cells has made significant progress in terms of yield, with an original efficiency of 5–10% to over 80% with current protocols [9–11]. Original protocols were based on random differentiation of three-dimensional pluripotent stem cell aggregates, known as embryoid bodies (EBs), to cardiomyocytes (CMs). Beating EBs are typically observed in the 7–10 days after starting the differentiation. Different techniques have been proposed to further purify cardiomyocytes from these aggregates including flow cytometry with selective markers such as SIRPA [12]. Even if multiple protocols for cardiac differentiation have been proposed, there is a current consensus to use monolayer-based protocols combined with glucose-deprived media with abundant lactates in order to confer cardiomyocytes a survival advantage over other cell types [9]. In addition, small molecules have progressively replaced cytokines and growth factors as these compounds revealed unstable and expensive to produce. Small molecules targeting the BMP and the WNT pathways have been shown as critical in inducing mesoderm induction, cardiomyocytes differentiation as well as cardiac specification to ventricular cardiomyocytes as opposed to atrial or junctional cardiomyocytes [5, 13, 14]. Indeed, most of the cardiomyopathies affect the ventricular cavities, especially the left ventricle which is the critical cavity for pumping the blood all over the body. The remarkable advances in cardiac differentiation protocols combined with high-throughput culture and phenotyping techniques have thus greatly enhanced the potential of iPSC-CMs for disease modeling and clinical applications.

5.2 Modeling Genetic Cardiac Arrhythmias

Multiple studies using hiPSC-CMs to model cardiovascular diseases initially focused on arrhythmic syndromes caused by channelopathies because of their clear, distinct, and easily assessable phenotypes both clinically and in cell culture. The channelopathies refer to diseases caused by the disturbed function of ion channel proteins expressed in the plasma membrane or of proteins that are involved in the regulation of expression or function of these ion channels. Ion channels expressed in the cardiomyocytes plasma membrane are critical for allowing the propagation of the cardiac action potential, which further triggers cardiac mechanical contraction. The cardiac action potential generates from the coordinated action of different ion channels for sodium (Na+), calcium (Ca2+), and potassium (K+) as illustrated in Fig. 5.1. When genetic abnormalities cause these channels to be dysfunctional, the resulting action potential is disturbed which will predispose to the occurrence of life-threatening cardiac arrhythmias.

Of these, congenital long QT syndrome (LQTS), catecholaminergic polymorphic ventricular tachycardia (CPVT), and Brugada syndrome (BrS) were the most commonly modeled diseases with iPSC as listed in Table 5.2 [52]. The majority of these genetic diseases are autosomal dominant with however variable expressivity. Channelopathies are typically associated with abnormalities on the electrocardiogram, which reflects action potential activity within the whole heart.

5.2.1 *Congenital Long QT Syndrome*

The long QT syndrome (LQTS) is a rare congenital disorder, with an estimated prevalence of 1 in 2000 individuals, which is characterized by a delayed repolarization of the heart following each heartbeat. This increases the risk of a particular form of polymorphic ventricular tachycardia called "torsades de pointes" (referring for its twisting appearance). These episodes can degenerate into more advanced cardiac

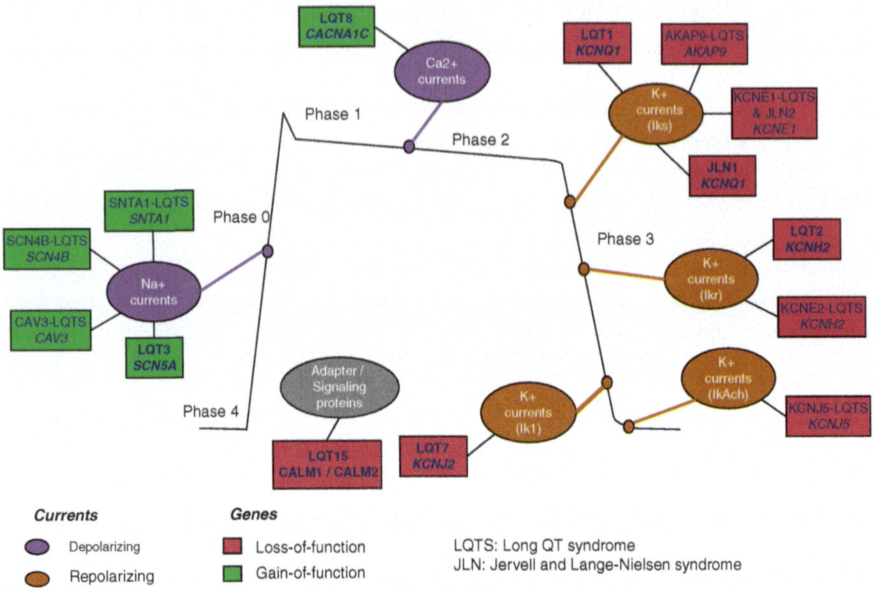

Fig. 5.1 Congenital long QT syndrome (LQTS): main currents, channels, and genes involved in the different types of long QT syndromes. The Jervell and Lange-Nielsen syndrome is a particular form of LQTS associated with severe bilateral sensorineural hearing loss. In bold, forms of LQTS that have been modeled using iPSCs

arrhythmias eventually leading to cardiac sudden death. LQTS is typically diagnosed in a familial context of sudden death, symptoms of fainting or syncope, and a prolongation of the QT interval on the ECG. LQTS is known to be caused by mutations in genes encoding for cardiac potassium, sodium, or calcium ion channels. To date, 15 genes have been identified and are associated with different types of LQTS. Among these, five types have been reproduced using patient-specific iPSCs (Table 5.2). In addition, LQTS can appear as an isolated disorder (called Romano-Ward syndrome with six types) or be associated with other defects; on top of long QT, the Andersen-Tawil syndrome (LQT7) associates developmental abnormalities and muscle weakness; the Timothy syndrome (LQT8) is characterized by subcutaneous syndactyly with abnormal facial features and signs of autism. Finally, the Jervell and Lange-Nielsen syndrome is characterized by a profound hearing loss from birth.

5.2.1.1 Long QT Type 1

The long QT syndrome type 1 (LQT1) represents 45% of the cases of LQTS and is caused by mutations affecting *KCNQ1* that encodes for part of the I_{Ks} slowly deactivating, delayed rectifier potassium channel (Fig. 5.1). More than 170 mutations

Table 5.2 List of channelopathies modeled with iPSCs

Disease	Locus	Phenotype	Gene correction	Ref
Congenital long QT syndrome (LQT)				
LQT1	*KCNQ1*	Prolonged action potential duration, aberrant potassium current	No	[15–18]
LQT2	*KCNH2*	Prolonged action potential duration, reduced potassium current IKr density	Yes	[17–26]
LQT3	*SCN5A*	Prolonged action potential duration, defective Na+ channel	Yes	[27–31]
LQT7 (Andersen-Tawil syndrome)	*KCNJ2*	Inhibition of endogenous IK1 by blocking the export of the channel	No	[32, 33]
LQT8 (Timothy syndrome)	*CACNA1C*	Irregular electrical activity, abnormal Ca2+ influx	No	[34]
LQT15	*CALM1*, *CALM2*	Prolonged action potential duration, impaired inactivation of L-type Ca2+ channels	Yes	[35–38]
Catecholaminergic polymorphic ventricular tachycardia (CPVT)				
CPVT1	*RYR2*	Sarcoplasmic reticulum calcium leakage, arrhythmias with delayed after depolarization	No	[39–47]
CPVT2	*CASQ2*	Increased diastolic calcium, arrhythmias with delayed after depolarization	No	[44, 48, 49]
Brugada syndrome				
BrS	*SNC5A*	Blunted inward sodium current, abnormal calcium handling	Yes	[50, 51]

(most missense) of this gene have been reported, but only a small number have been studied in iPSC [15–18]. Overall, these mutations result in decreased outward potassium current creating a delay in ventricular repolarization and with prolonged duration of action potential in iPSC-derived cardiomyocytes.

5.2.1.2 Long QT Type 2

The long QT syndrome type 2 (LQT2) represents another 45% of the cases of LQTS and is caused by mutations affecting *KCNH2* that encodes for part of the I_{Kr} rapidly deactivating, delayed rectifier potassium channel (Fig. 5.1). More than 200 *KCNH2* mutations have been detected and typically cause rapid closure of the potassium channel, a decrease in the normal rise of I_{Kr} and consequent prolongation of the action potential duration. Because of its prevalence, multiple studies have reported LQT2 modeling using iPSCs [17–26]. Different studies are also using LQT2 iPSC-derived cardiomyocytes to better understand the regulation of hERG channel function [25] or expression [20].

5.2.1.3 Long QT Type 3

The long QT syndrome type 3 (LQT3) represents 7% of the cases of LQTS. LQT3 is caused by mutations in *SCN5A* encoding for the sodium channel. Gain-of-function mutations cause persistent inward sodium current in the plateau phase, which contributes to prolonged repolarization. Interestingly, some loss-of-function mutations have been reported in the same gene, but they lead to different forms of cardiac arrhythmias, including the Brugada syndrome. LQT3 iPSC-derived cardiomyocytes recapitulate the electrophysiological features of the disease as well as the therapeutic efficacy of sodium channel blockers [27–30].

5.2.1.4 Long QT Type 7 or Andersen-Tawil Syndrome

The long QT syndrome type 7 (LQT7) is a rare form of LQTS, caused by mutations in *KCNJ2*, a gene coding for potassium channel 2 protein which is expressed in the heart but also the skeletal muscle further explaining the extra-cardiac abnormalities observed in the Andersen-Tawil syndrome. Modeling studies using iPSC-CM have focused on the electrophysiological consequence of the dysfunctional channel as found in LQT7 patients [32, 33]. Here again, the defective potassium channel translates into a deficient repolarization phase, prolonged action potential, and higher risk of developing arrhythmias. Additional experiments also revealed irregular calcium handling in LQT7 iPS-derived cardiomyocytes, probably through a compensatory activation of the reverse mode of the Na+/Ca2+ exchanger (NCX). Consequently, flecainide (a NCX inhibitor) was able to suppress arrhythmic events observed in the LQT7 iPSC-derived cardiomyocytes [33]. Flecainide was however also suggested to increase the potassium currents carried by the KCNJ2 channels [53], which could further support its efficacy in LQT7 patients.

5.2.1.5 Long QT Type 8 or Timothy Syndrome

The long QT syndrome type 8 (LQT8) is another rare form of LQTS, caused by gain-of-function mutations in *CACNA1C*, the gene coding for the L-type calcium currents. As for the Andersen-Tawil syndrome, the Timothy syndrome associates extra-cardiac abnormalities, especially at the neuronal level. In one study [34], electrophysiological recording and calcium imaging studies of LQT8 iPSC-derived cardiomyocytes revealed excess calcium influx, prolonged action potentials and irregular electrical activity but also irregular contraction, and abnormal calcium transients which fits with the essential role of calcium in the excitation-contraction coupling of cardiomyocytes. The use of a voltage-dependent calcium channel inhibitor was able to rescue the electrical and calcium signaling properties in these cells [34]. Interestingly, alterations of calcium signaling were also found in Timothy syndrome iPSC-derived neurons [54, 55].

5.2.1.6 Long QT Type 15 or Calmodulinopathies

The long QT syndrome type 15 (LQT15) is a rare form of LQTS that is not linked to mutations in genes coding for the cardiac ion channels per se but in the *CALM1/CALM2* genes coding for calmodulin, a calcium-binding protein that regulates a wide range of biological processes including calcium-dependent inactivation of L-type calcium currents [56]. The first human *CALM* disease mutations were only recently discovered and were associated with a range of ventricular arrhythmia syndromes including long QT syndrome (LQT15). Four different studies have explored LQT15 iPSC-derived cardiomyocytes [35–38], found prolonged action potentials in these cells, and confirmed the abnormal Ica inactivation as predicted in animal studies. In addition, genetic interference strategy of the mutated allele was sufficient to nearly normalize electrophysiological abnormalities found in LQT15 iPSC-derived cardiomyocytes [35, 38].

5.2.1.7 Jervell and Lange-Nielsen Syndrome

In contrast to other types of long QT that appear as autosomal dominant traits, the JLN syndrome is a recessive trait resulting from homozygous or compound heterozygous mutations in *KCNQ1* or *KCNE1*. These two genes encode subunits of a channel conducting the slow component of the delayed rectifier potassium current (called I_{Ks}) as shown in Fig. 5.1. In one study [57], patient-specific iPSC were generated from a patient with JLNS-causing mutations in *KCNQ1* and were differentiated into cardiomyocytes. Electrophysiological studies showed the severe reduction or absence of I_{Ks} resulting in pronounced action potential prolongation. Mutations were further shown as directly affecting the channel function or as reducing channel trafficking at the membrane in a gene-dosage dependent fashion [57].

5.2.2 *Catecholaminergic Polymorphic Ventricular Tachycardia*

Catecholaminergic polymorphic ventricular tachycardia (CPVT) is an inherited arrhythmia often leading to sudden cardiac death in children and young adults and occurring with an estimated prevalence of 1 in 10,000 individuals. CPVT is characterized by polymorphic/bidirectional ventricular tachycardia induced by adrenergic stimulation associated with emotional stress or with physical exercise. There are two forms of CPVT. CPVT1 (50–60% of CPVT cases) is caused by mutations in *RYR2*, encoding for ryanodine receptor type 2. CPVT1 is inherited by a dominant mechanism. CPVT2 is caused by mutations in *CASQ2*, encoding for cardiac calsequestrin 2 and is inherited by recessive mechanism. Both RYR2 and CASQ2 proteins are major components of a complex that controls coordinated calcium release from the sarcoplasmic reticulum during the systole (i.e., the contraction phase).

Analysis of CPVT1 and CPVT2 patient-specific iPSC-derived cardiomyocytes globally revealed the need of a beta-adrenergic stimulation to induce arrhythmic events, while there is an absence of baseline arrhythmic events in these cells [39–49]. This feature is consistent with the clinical phenotype. Additional characterization of CPVT1/CPVT2 iPSC-CM revealed underlying abnormalities in cardiomyocytes ultrastructure [44, 49] as well as calcium handling with increased diastolic calcium levels in line with an increased RYR-mediated calcium leak from the sarcoplasmic reticulum [39–44, 46, 47, 49].

5.2.3 Brugada Syndrome

Brugada syndrome is another genetic condition associated with an increased risk of sudden cardiac death. Approximately 20% of the cases are caused by mutations in *SCN5A* that encodes for the sodium ion channel (Fig. 5.1). As explained earlier, *SCN5A* is also involved in long QT syndrome type 3 (LQT3), and the Brugada syndrome is often referred as an overlap syndrome [58]. In addition, the majority of Brugada syndrome remains genetically elusive. Even if the description of the underlying mechanisms is incomplete, features of Brugada syndrome were observed in iPSC-cardiomyocytes from patients with *SCN5A* mutations [50, 51]. There was a reduction in the sodium current density that favors the occurrence of arrhythmias [50, 51]. An additional impact on calcium handling in these cells was also observed [50]. In contrast, iPSC-cardiomyocytes derived from genotype-negative Brugada syndrome did not exhibit a clear cellular phenotype suggesting a different mechanism for these patients which not be clearly modeled using iPSCs [59].

5.3 Modeling Genetic Cardiomyopathies

Modeling of cardiomyopathies using hiPSC-derived cardiomyocytes has essentially been applied for familial cardiomyopathies as a disease-causing gene mutation is usually described or suspected. Most cardiomyopathies are characterized by disrupted sarcomeric alignment, a phenomenon also called myocardial disarray that further supports myocardial dysfunction.

Sarcomeric cardiomyopathies refer to cardiac disorders associated with a primary defect (i.e., a mutant protein) in the sarcomere, the basic contractile unit of a striated muscle tissue. Sarcomeres represent the microanatomical unit that is repeated to form myofibrils. The sarcomere is made of aligned myofilaments with a typical length of about 1.6–2.2 μm in humans. Multiple proteins assemble to form a competent sarcomere (Fig. 5.2a–b). In addition, if the sarcomere represents the engine of cardiomyocytes, the beat-to-beat contraction is activated through

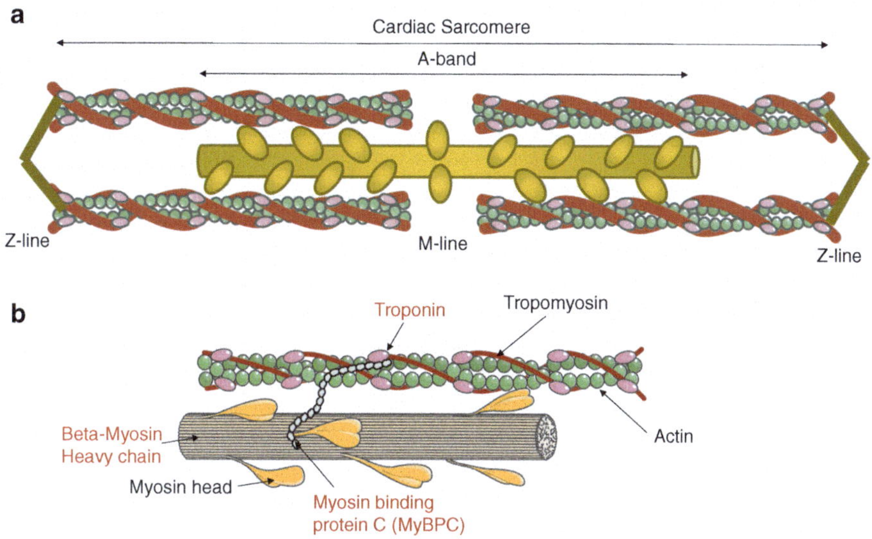

Fig. 5.2 (**a**) Schematic representation of a cardiac sarcomere. (**b**) Composition of a sarcomere: actin filaments (thin filaments) and myosin filaments (thick filaments) are the major components. The action molecules are covered with tropomyosin and troponin molecules that will allow bridging to myosin heads in the presence of calcium. This interaction allows a sliding mechanism between the two filaments, which creates the contraction. In red, molecules that were investigated in iPSC

organized fluctuations of calcium from the stores (in the sarcoplasmic reticulum) to the sarcomere. A rise in the cytosolic level of calcium is essential for the sarcomere to contract, but a reuptake of this calcium back to the sarcoplasmic reticulum is essential to allow sarcomere relaxation and therefore appropriate preparation for the next contraction phase. This calcium movement is made possible by an organized network of calcium handling proteins (Fig. 5.3).

Mutations affecting sarcomere proteins, calcium handling proteins, or key regulatory molecules (that are still progressively discovered) can cause some form of cardiomyopathies, mainly associated with cardiac dilated cardiomyopathy (DCM) or hypertrophic cardiomyopathy (HCM). In addition, the expressivity of these mutations is heterogeneous, and mutations affecting the same gene can either be associated with DCM or HCM, probably depending on the myofilament tension [60], thus urging for better physiopathological understanding in appropriate models. Overall, a small number of these mutations have been studied using iPSC (Table 5.3) potentially as modeling these forms of cardiomyopathies integrates a challenging architectural level to allow proper assembly and alignment of cardiac structures.

Finally, familial cardiomyopathies modeled with iPSC also include arrhythmogenic right ventricular cardiomyopathy (ARVC), a disease characterized by fatty infiltration of the ventricles with predominance on the right ventricle, cardiac arrhythmias, and right ventricular dysfunction [99].

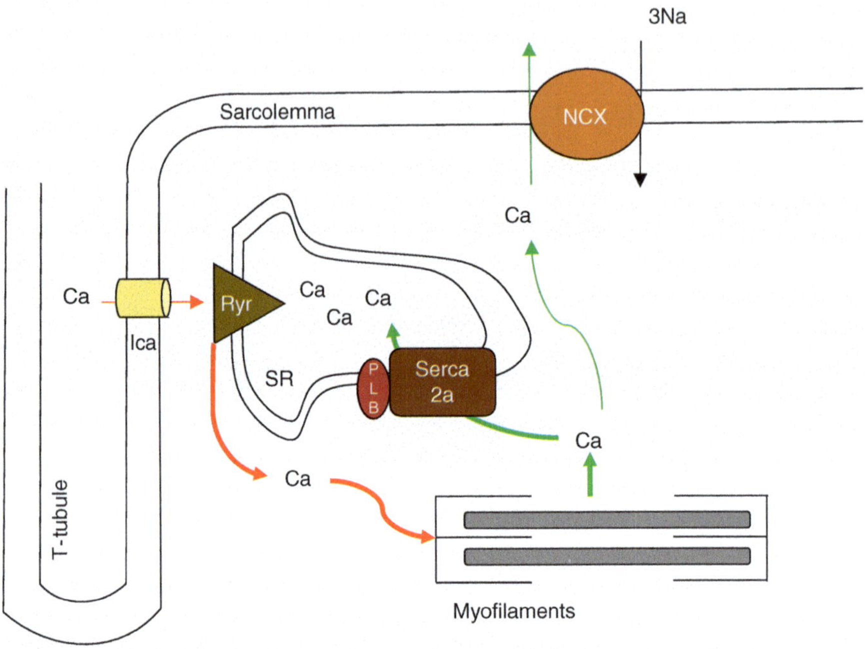

Fig. 5.3 Basis of the excitation-contraction coupling. At each depolarization of the plasma membrane, voltage-dependent calcium channels (Ica) open, and the small amount of calcium entering the cell triggers the release of a larger amount of calcium from the sarcoplasmic reticulum (SR) via ryanodine receptors (Ryrs). The rise in the systolic calcium level supports the myofilament contraction. During the relaxation phase, calcium reuptake to the SR occurs through the SR Ca2+-ATPase (Serca2a) active pump. Phospholamban (PLB) acts as a primary regulator of Serca2a activity. Calcium levels are finally adjusted through the sodium calcium exchanger (NCX)

Table 5.3 List of structural cardiomyopathies modeled with iPSC

Disease	Locus	Phenotype	Gene correction	Ref
Dilated cardiomyopathy				
Troponin	*TNNT2*	Decreased contractility, altered calcium regulation	No	[60–63]
Titin	*TTN*	Sarcomere abnormalities, impaired contractility	Yes	[64]
Phospholamban	*PLN*	Unstable calcium handling, impaired contractility, and arrhythmic events	Yes	[65, 66]
Lamin	*LMNA*	Accelerated senescence, stress-induced apoptosis	No	[67, 68]

Table 5.3 (continued)

Disease	Locus	Phenotype	Gene correction	Ref
	BAG3	Disrupted myofibril structure, reduced protein expression, impaired contractility	Yes	[69]
	RBM20	Defective calcium handling, sarcomere disarray	No	[70–72]
Desmin	*DES*	Abnormal sarcomere architecture	No	[73]
Hypertrophic cardiomyopathy				
	MYH7	Increased cell size, sarcomere disarray, arrhythmic events, calcium handling abnormalities	No	[74, 75]
	MYBPC3	Cell hypertrophy, myofibrillar disarray, calcium handling abnormalities	No	[76–79]
	PRKAG2	Cell hypertrophy, triggered arrhythmias	Yes	[80]
Cardio-facio-cutaneous syndrome	*BRAF*	Cell hypertrophy mediated by abnormal RAS/MAPK signaling Increased contraction	No	[81, 82]
Leopard syndrome	*PTPN11*	Cell hypertrophy, higher expression of hypertrophic genes	No	[83]
Friedreich's ataxia	*FXN* (frataxin)	Disorganized mitochondrial network, impaired iron homeostasis, aberrant calcium signaling	No	[84, 85]
Other cardiomyopathies				
Arrhythmogenic right ventricular cardiomyopathy/dysplasia (ARVC/D)	*PKP2*	Desmosomal distortion, exaggerated lipogenesis, increased apoptosis	No	[86–88]
Duchenne muscular dystrophy	*DMD*	Dystrophin deficiency, impaired contractility, mitochondrial damage, and increased apoptosis	No	[89]
Metabolic cardiomyopathy				
Diabetic cardiomyopathy		Changes in metabolic genes expression	No	[90]
Barth syndrome	*TAZ*	Increased ROS, mitochondrial dysfunction with decreased respiration	Yes	[91, 92]
Pompe disease	*GAA*	Glycogen accumulation	No	[93–96]
Fabry disease	*GLA*	Gb3 accumulation in lysosomes	No	[97, 98]

5.3.1 *Dilated Cardiomyopathy (DCM)*

Patient-specific iPSC-derived cardiomyocytes carrying mutations in different DCM-related genes (i.e., *TNNT2, TTN, PLN, LMNA, DES, BAG3, RBM20*) have shown characteristic sarcomeric disarray associated with a reduction in contractile force. Different pathophysiological mechanisms have however been reported, depending on the role of mutated protein.

5.3.1.1 Sarcomeric Proteins

Cardiac troponin T mutation (*TNNT2*) was the first explored using iPSC [60–63]. In line with its essential role in the sarcomere organization, mutated iPSC-derived cardiomyocytes show ultrastructural abnormalities with scattered distribution of Z lines. In addition, these cells exhibited impaired calcium handling, decreased tolerance to beta-adrenergic challenge, and reduced contractile force [62]. It was further demonstrated that the mutated TNNT2 caused the cardiomyopathy by nuclear translocation and epigenetic activation of PDE2A and PDE3A genes that leads to cardiac dysfunction by blunting the inotropic cyclic AMP pathway. Pharmacological inhibition of PDE2A and PDE3A attenuated the phenotype [63] as well as the use of a myosin activator [61].

In another study [64], DCM caused by mutations that truncate the sarcomere protein titin (TTN) was modeled using iPSC-derived engineered cardiac microtissues. This technology offers the advantage to induce a better maturation of cardiomyocytes and to offer a more suitable platform to measure contractility and relaxation parameters as compared to cells in monolayers. In this study, the sarcomere disarray and the impaired contractility were reproduced [64]. In addition, different types of mutations were modeled with different levels of cardiac dysfunction in line with clinical observations. This indicated that iPSC technology can also be used for prognostication of titin-related DCMs.

5.3.1.2 Calcium Handling Proteins

A mutation in the coding region of the phospholamban (PLN) gene (R14del) was modeled in iPSC [65, 66]. Heterozygous patients exhibit left ventricular dilation and ventricular arrhythmias. At the cell level, patient-specific iPSC-CMs exhibit abnormal calcium cycling with electrical instability and arrhythmic events [65]. The mutation was associated with an abnormal distribution of PLN protein. Using 3D-engineered cardiac tissues, it was further shown that the mutation directly impairs contractility [66]. These phenotypic abnormalities were reversed after gene correction [65, 66].

5.3.1.3 Other Proteins

In separate studies, iPSC-CMs carrying DCM-associated mutations in lamin A/C (*LMNA*) [67, 68], the co-chaperone BCL2-associated athanogene 3 (*BAG3*) [69], the spliceosome RNA-binding motif protein 20 (*RBM20*) [70–72], or the desmin (*DES*) [73] genes also demonstrated typical morphological and functional abnormalities.

5.3.2 Hypertrophic Cardiomyopathy (HCM)

HCM is characterized by an asymmetrical left ventricular hypertrophy which results in higher risk of life-threatening ventricular arrhythmias and higher risk of heart failure. About 70% of HCM are explained by mutations affecting either the beta-myosin heavy chain (*MYH7*, 35%) or the cardiac myosin-binding protein C (*MYBPC3*, 35%), two essential components of the sarcomere (Fig. 5.2b). iPSC-CMs carrying HCM-associated mutations in both genes were generated and found to recapitulate some features of the disease, including increased cell size, sarcomere disarray, and arrhythmic events [74–79]. Abnormal calcium handling and intracellular calcium elevation appeared as key pathological mechanisms that can be improved by the calcium channel blocker verapamil [75]. Gene replacement in iPSC-CM carrying a *MYBPC3* mutation was also associated with a partial restoration of HCM phenotype [78].

HCM can also be observed in dysmorphogenetic disorders, such as the cardio-facio-cutaneous syndrome or the Leopard syndrome, or in genetic condition such as Friedreich's ataxia. HCM associated with these conditions was modeled using iPSC technology [81–84]. HCM-associated mutations in *BRAF* were investigated in 3D-engineered tissues, showing a hypertrophic phenotype with increased tissue size, contraction, and relaxation rates [81].

5.3.3 Other Cardiomyopathies

Patient-specific iPSC-CMs carrying mutations in the desmosomal protein plakophilin-2 (*PKP2*) were generated to model arrhythmogenic right ventricular cardiomyopathy [86–88]. These cells exhibited some of the features observed in ARVC including desmosome disorganization and increased lipogenesis and apoptosis but after stimulation with a lipogenic cocktail [87]. Modeling ARVC is however challenging because of its adult-onset phenotype and potential triggers of unknown nature.

5.4 Modeling Metabolic Cardiomyopathy

Metabolic cardiomyopathies can be caused by different disturbances in metabolism, among which diabetes mellitus is the most prevalent. Diabetes can indeed change the metabolism of fatty acids, the main source of energy in the heart. The inefficient energy production leads to an impaired excitation-contraction coupling that favors cardiac dysfunction. However, metabolic cardiomyopathy can also appear as a manifestation of systemic metabolic diseases caused by deficiencies of various enzymes in a variety of metabolic pathways. Even if metabolic disorders account for a minority of cardiomyopathy cases, defects in mitochondrial oxidative phosphorylation (Barth syndrome), glycogen-related disorder (Pompe disease), and lysosomal diseases (Fabry disease) are frequently associated with cardiac findings and were modeled using iPSC (Table 5.3).

5.4.1 Diabetic Cardiomyopathy

In a single study, iPSC-derived cardiomyocytes were placed for a prolonged time in a diabetic milieu consisting of a combination of glucose, ET-1, and cortisol [90]. Transcriptomic profiling revealed consistent changes in metabolic genes and suggested that this chemical stimulation can induce a phenotypic surrogate of diabetic cardiomyopathy.

5.4.2 Barth Syndrome

Barth syndrome (BTHS) is a mitochondrial disorder caused by mutation in the tafazzin (TAZ) gene, encoding for a key mitochondrial protein involved in altering a fat called cardiolipin. Barth syndrome is characterized by the presence of a dilated cardiomyopathy with weakness in skeletal muscles. Investigations in cells and engineered tissues using BTHS iPSC-derived cardiomyocytes revealed typical metabolic (excess accumulation of ROS), structural (irregular sarcomeres), and functional (decreased contractility) abnormalities [91, 92].

5.4.3 Pompe Disease

Pompe disease is an autosomal recessive lysosomal storage disease caused by mutations in *GAA*, the gene coding for the glycolytic enzyme α-glucosidase, which lead to glycogen accumulation in myocytes as well as in iPSC-derived cardiomyocytes [94–96]. Using the iPSC model, mass spectrometry analysis of *N*-linked glycans further revealed that Pompe cardiomyopathy could be caused by a glycan-processing abnormality [94].

5.4.4 *Fabry Disease*

Fabry disease is an X-linked recessive lysosomal storage disease that is caused by mutation in *GLA*, resulting in a deficient or absent activity of the enzyme α-galactosidase A. Consequently, there is a progressive lysosomal accumulation of globotriaosylceramide (Gb3) in a variety of cell types including cardiomyocytes. This Gb3 accumulation was found in the lysosomes of Fabry iPSC-derived cardiomyocytes [97, 98]. It was further demonstrated that substrate reduction therapy via glucosylceramide synthase inhibition (ibiglustat) was able to prevent accumulation and to clear lysosomal GL-3 in cardiomyocytes [98].

5.5 Modeling Drug Responses in iPSC-Derived Cardiomyocytes

5.5.1 *Pharmacological Testing in Diseased iPSC-Derived Cardiomyocytes*

Different studies have used patient-specific iPSC-derived cardiomyocytes as a platform for pharmacological testing of available or novel drugs, especially in studies on channelopathies. For instance, the use of calcium channel blockers was shown protective in LQT1 and LQT2 [18, 22], as assessed by significant reduction in action potential duration and cellular arrhythmogenic events. In one study [31], it was further proposed to adjust therapeutic interventions of the clinical level according to drug actions observed in iPSC-derived cardiomyocytes. The patient presented with a complex genetic form of LQTS associating mutations in two major cardiac ion channels (i.e., *SCN5A* affecting the NA+ currents and *KCNH2* affecting the potassium I_{Kr} current). Investigations in patient-specific iPSC-derived cardiomyocytes revealed a predominant role of the late Na+ channel current in the cellular phenotype, which was consequently improved by the use of a specific Na+ channel blocker. Administration of this Na+ channel blocker to the patient resulted in an improved management of arrhythmias. This study exemplified the use of iPSC technology to guide treatments in a given patient, a scenario that however remains unusual in clinical practice.

On the other hand, iPSC-derived cardiomyocytes revealed as a powerful tool to identify therapeutic actions of compounds with novel mechanisms of action. In a first study using LQT1 patient iPSC-CM, a novel small molecule designed as affecting potassium channel activator (called ML277) was shown to reverse the decreased I_{Ks} potassium current and to partially normalize action potential duration, thereby indicating a therapeutic potential in LQT1 patients [15]. Similarly in another study, a novel small molecular (LUF7346) acting as a hERG/iKr allosteric modulator was tested on a library of isogenically matched, diseased, and genetically engineered iPSC from normal, LQT1 and LQT2 patients [17]. Even if hERG

(*KNCH2*) is primarily involved in LQT2, the hERG allosteric modulator was found to rescue all of the conditions, thus further showing the interdependence of the different cardiac currents.

Patient-specific iPSC-derived cardiomyocytes can also be used to further demonstrate unanticipated anti-arrhythmic effects of some drugs designed for another purpose. In a very elegant study [100], lumacaftor (a drug acting as a chaperone during protein folding and initially developed to treat cystic fibrosis by enhancing CFTR channel trafficking to the membrane) was shown to have anti-arrhythmic effect in iPSC-CM from patients with LQT2. Interestingly, the authors used a panel of iPSC from five patients with two different classes of KCNH2 mutations: some of the mutations (class 1) were previously found to disrupt the channel synthesis, while some other mutations (class 2) affected the channel intracellular transport to the cell membrane. By enhancing hERG trafficking to the cell membrane, lumacaftor was able to improve cardiac repolarization in iPSC-CM from patients with class 2 mutations. However, an additional and unanticipated action on calcium handling in all types of iPSC-derived cardiomyocytes was also observed, thus supporting a potential therapeutic action all LQT2 patients and particularly those with class 2 mutations [100].

Similarly, different pharmacological testings were performed on CPVT iPSC-derived cardiomyocytes with the objective of fixing the abnormal calcium leakage at the ryanodine receptor level. Dantrolene, a commercially available muscle relaxant with inhibitory action on RYR2, was thus proved effective in reducing aberrant calcium release from the sarcoplasmic reticulum, thereby reducing the arrhythmic events in these cells [42]. S107 is benzothiazepine derivative with a mode of action that consists of stabilizing the closed state of the ryanodine receptor 2. Preincubation of CPVT iPSC-derived cardiomyocytes significantly decreased the occurrence of arrhythmias in response to beta-adrenergic stimulation [46].

5.5.2 Cardiovascular Pharmacogenomics

Another question that can be addressed using the iPSC technology is modeling inter-patient variability in the response to a given drug. This includes cardiotoxic effects that usually develop in a limited number of subjects with a predisposing genetic background [101]. In contrast to studies modeling familial cardiomyopathies, the exact genetic determinants of this particular drug cytotoxic response are most of the time unknown (i.e., probably based on a the additive effect of different genetic polymorphisms), but combining pharmacology, genomics, and iPSC technology has proven helpful in reproducing drug-induced cardiovascular side effects in vitro. A first example is drug-induced long QT, a pharmacogenomic syndrome that mimics congenital long QT but is primarily induced by drugs in apparently healthy subjects [102]. It is thought that these individuals have a limited cardiac repolarization reserve. In a prospective study in 96 subjects, iPSC-derived cardiomyocytes were generated from patients presenting in vivo with extremely low

($n = 10$) or high ($n = 10$) changes in cardiac repolarization in response to a pharmacological challenge with sotalol (a hERG blocker) [103]. In vitro, the responses to sotalol were highly variable but strongly correlated with the inter-individual differences observed in vivo. This illustrates the potential of using a genetically diverse panel of subject-specific iPSCs to model complex and acquired phenotypes. In addition, transcriptomic profiling of these cells identified significant changes in the expression of genes involved in downstream regulation of cardiac repolarization machinery [103]. This proposes a new mechanism underlying high sensitivity to sotalol, but whether this translates to other drugs inducing long QT remains to be proven.

Similarly, patient-specific genetic variations have been shown to affect the occurrence of doxorubicin-induced cardiotoxicity. Doxorubicin is an antineoplastic drug commonly used in other chemotherapy agents in patients with breast cancer. Using a library of iPSCs from female patients who experienced doxorubicin-induced heart failure, it was shown that the predilection to develop doxorubicin-induced cardiotoxicity can be reproduced in iPSC-CMs with decreased cell viability and mitochondrial and metabolic function when exposed to doxorubicin as compared to control iPSC-CMs [104]. An increased reactive oxygen species (ROS) production was observed in the predisposed iPSC-CMs.

These two studies illustrate the potential of iPSCs technology for the prediction of individual risk and for its use in precision medicine. However, it remains to be demonstrated that this approach can be applied prospectively to adjust medications according to an individual's profile of predicted reactions using its iPSC.

5.6 Future Developments and Conclusions

Collectively, multiple studies have now demonstrated that patient-specific iPSC-CMs can serve as disease modeling of cardiomyopathies and drug discovery platforms. Diseased hiPSC-CMs can recapitulate some clinical phenotypes, including arrhythmias and contractile dysfunction, as observed in the patients from which the cells were derived. This was particularly developed in the context of monogenic cardiomyopathies, but more recent studies also suggest that iPSC-CMs can also be used to assess the unique interactions between individual patients' genetic backgrounds and environmental or pharmacological factors.

However, it is likely that the full potential of iPSC for modeling of cardiac disorders has still not been reached. First, the iPSC-derived cardiomyocytes are functionally immature as compared to their adult mature counterparts [105]. One of the major challenges hindering full realization of iPSC-derived myocytes thus relates to uncertainties in the maturation status of the myocytes (at the biochemical, electrophysiological, and mechanical levels) and future improvements in this field might revealed extremely important to achieve better disease modeling. The combination of iPSC technology with bioengineering techniques is an encouraging approach by improving iPSC-CMs maturation and providing a more suitable platform for the measurement of cardiac functional parameters (including contraction and relaxation) [66, 106–108]. Second, there is diversity of endpoints that have still not been fully investigated in iPSC-CMs. Most of the current studies have primarily focused on electrophysiological and architectural characterization of the generated cells, but the description of other key determinants of cardiomyocytes function is only starting. This notably includes mitochondrial and metabolic activities or relaxation and cell stiffness. In addition, there is a need for developing standardized protocols mimicking the cardiac pathological environment. For instance, exploring the response of iPSC-CMs in different conditions of mechanical stretch has only been recently proposed, whereas the mechanosensitivity plays a critical role in the development of cardiomyopathies [60, 109]. All of these future developments will further push iPSC into a more physiological cardio-mimicking model.

References

1. Report of the WHO/ISFC task force on the definition and classification of cardiomyopathies. Br Heart J. 1980;44:672–3.
2. Elliott P, et al. Classification of the cardiomyopathies: a position statement from the European Society Of Cardiology Working Group on Myocardial and Pericardial Diseases. Eur Heart J. 2008;29:270–6.
3. Chen IY, Matsa E, Wu JC. Induced pluripotent stem cells: at the heart of cardiovascular precision medicine. Nat Rev Cardiol. 2016;13:333–49.
4. Jeziorowska D, Korniat A, Salem JE, Fish K, Hulot JS. Generating patient-specific induced pluripotent stem cells-derived cardiomyocytes for the treatment of cardiac diseases. Expert Opin Biol Ther. 2015;15:1399–409.
5. Matsa E, Ahrens JH, Wu JC. Human induced pluripotent stem cells as a platform for personalized and precision cardiovascular medicine. Physiol Rev. 2016;96:1093–126.
6. Shi Y, Inoue H, Wu JC, Yamanaka S. Induced pluripotent stem cell technology: a decade of progress. Nat Rev Drug Discov. 2017;16:115–30.
7. Kattman SJ, Huber TL, Keller GM. Multipotent flk-1+ cardiovascular progenitor cells give rise to the cardiomyocyte, endothelial, and vascular smooth muscle lineages. Dev Cell. 2006;11:723–32.
8. Kouskoff V, Lacaud G, Schwantz S, Fehling HJ, Keller G. Sequential development of hematopoietic and cardiac mesoderm during embryonic stem cell differentiation. Proc Natl Acad Sci U S A. 2005;102:13170–5.
9. Burridge PW, Holmstrom A, Wu JC. Chemically defined culture and cardiomyocyte differentiation of human pluripotent stem cells. Curr Protoc Hum Genet. 2015;87:21.3.1–15.
10. Burridge PW, et al. Chemically defined generation of human cardiomyocytes. Nat Methods. 2014;11:855–60.
11. Jeziorowska D, et al. Differential sarcomere and electrophysiological maturation of human iPSC-derived cardiac myocytes in monolayer vs. aggregation-based differentiation protocols. Int J Mol Sci. 2017;18:E1173.
12. Dubois NC, et al. SIRPA is a specific cell-surface marker for isolating cardiomyocytes derived from human pluripotent stem cells. Nat Biotechnol. 2011;29:1011–8.
13. Karakikes I, et al. Small molecule-mediated directed differentiation of human embryonic stem cells toward ventricular cardiomyocytes. Stem Cells Transl Med. 2014;3:18–31.
14. Mazzotta S, Neves C, Bonner RJ, Bernardo AS, Docherty K, Hoppler S. Distinctive roles of canonical and noncanonical Wnt signaling in human embryonic cardiomyocyte development. Stem Cell Reports. 2016;7:764–76.

15. Ma D, et al. Characterization of a novel KCNQ1 mutation for type 1 long QT syndrome and assessment of the therapeutic potential of a novel IKs activator using patient-specific induced pluripotent stem cell-derived cardiomyocytes. Stem Cell Res Ther. 2015;6:39.
16. Moretti A, et al. Patient-specific induced pluripotent stem-cell models for long-QT syndrome. N Engl J Med. 2010;363:1397–409.
17. Sala L, et al. A new hERG allosteric modulator rescues genetic and drug-induced long-QT syndrome phenotypes in cardiomyocytes from isogenic pairs of patient induced pluripotent stem cells. EMBO Mol Med. 2016;8:1065–81.
18. Wang Y, et al. Genome editing of isogenic human induced pluripotent stem cells recapitulates long QT phenotype for drug testing. J Am Coll Cardiol. 2014b;64:451–9.
19. Bellin M, et al. Isogenic human pluripotent stem cell pairs reveal the role of a KCNH2 mutation in long-QT syndrome. EMBO J. 2013;32:3161–75.
20. Caballero R, et al. Tbx20 controls the expression of the KCNH2 gene and of hERG channels. Proc Natl Acad Sci U S A. 2017;114:E416–25.
21. Duncan G, Firth K, George V, Hoang MD, Staniforth A, Smith G, Denning C. Drug-mediated shortening of action potentials in LQTS2 human induced pluripotent stem cell-derived cardiomyocytes. Stem Cells Dev. 2017;26(23):1695–705.
22. Itzhaki I, et al. Modelling the long QT syndrome with induced pluripotent stem cells. Nature. 2011;471:225–9.
23. Jouni M, et al. Toward personalized medicine: using cardiomyocytes differentiated from urine-derived pluripotent stem cells to recapitulate electrophysiological characteristics of type 2 long QT syndrome. J Am Heart Assoc. 2015;4:e002159.
24. Lahti AL, et al. Model for long QT syndrome type 2 using human iPS cells demonstrates arrhythmogenic characteristics in cell culture. Dis Model Mech. 2012;5:220–30.
25. Liu QN, Trudeau MC. Eag domains regulate LQT mutant hERG channels in human induced pluripotent stem cell-derived cardiomyocytes. PLoS One. 2015;10:e0123951.
26. Matsa E, Rajamohan D, Dick E, Young L, Mellor I, Staniforth A, Denning C. Drug evaluation in cardiomyocytes derived from human induced pluripotent stem cells carrying a long QT syndrome type 2 mutation. Eur Heart J. 2011;32:952–62.
27. Fatima A, et al. The disease-specific phenotype in cardiomyocytes derived from induced pluripotent stem cells of two long QT syndrome type 3 patients. PLoS One. 2013;8:e83005.
28. Ma D, et al. Modeling type 3 long QT syndrome with cardiomyocytes derived from patient-specific induced pluripotent stem cells. Int J Cardiol. 2013b;168:5277–86.
29. Malan D, et al. Human iPS cell model of type 3 long QT syndrome recapitulates drug-based phenotype correction. Basic Res Cardiol. 2016;111:14.
30. Paci M, Passini E, Severi S, Hyttinen J, Rodriguez B. Phenotypic variability in LQT3 human induced pluripotent stem cell-derived cardiomyocytes and their response to antiarrhythmic pharmacologic therapy: an in silico approach. Heart Rhythm. 2017;14(11):1704–12.
31. Terrenoire C, et al. Induced pluripotent stem cells used to reveal drug actions in a long QT syndrome family with complex genetics. J Gen Physiol. 2013;141:61–72.
32. Gelinas R, et al. Characterization of a human induced pluripotent stem cell-derived cardiomyocyte model for the study of variant pathogenicity: validation of a KCNJ2 mutation. Circ Cardiovasc Genet. 2017;10:e001755.
33. Kuroda Y, et al. Flecainide ameliorates arrhythmogenicity through NCX flux in Andersen-Tawil syndrome-iPS cell-derived cardiomyocytes. Biochem Biophys Rep. 2017;9:245–56.
34. Yazawa M, Hsueh B, Jia X, Pasca AM, Bernstein JA, Hallmayer J, Dolmetsch RE. Using induced pluripotent stem cells to investigate cardiac phenotypes in Timothy syndrome. Nature. 2011;471:230–4.
35. Limpitikul WB, et al. A precision medicine approach to the rescue of function on malignant calmodulinopathic long-QT syndrome. Circ Res. 2017;120:39–48.
36. Pipilas DC, et al. Novel calmodulin mutations associated with congenital long QT syndrome affect calcium current in human cardiomyocytes. Heart Rhythm. 2016;13:2012–9.

37. Rocchetti M, et al. Elucidating arrhythmogenic mechanisms of long-QT syndrome CALM1-F142L mutation in patient-specific induced pluripotent stem cell-derived cardiomyocytes. Cardiovasc Res. 2017;113:531–41.
38. Yamamoto Y, et al. Allele-specific ablation rescues electrophysiological abnormalities in a human iPS cell model of long-QT syndrome with a CALM2 mutation. Hum Mol Genet. 2017;26:1670–7.
39. Di Pasquale E, et al. CaMKII inhibition rectifies arrhythmic phenotype in a patient-specific model of catecholaminergic polymorphic ventricular tachycardia. Cell Death Dis. 2013;4:e843.
40. Fatima A, et al. In vitro modeling of ryanodine receptor 2 dysfunction using human induced pluripotent stem cells. Cell Physiol Biochem. 2011;28:579–92.
41. Itzhaki I, et al. Modeling of catecholaminergic polymorphic ventricular tachycardia with patient-specific human-induced pluripotent stem cells. J Am Coll Cardiol. 2012;60:990–1000.
42. Jung CB, et al. Dantrolene rescues arrhythmogenic RYR2 defect in a patient-specific stem cell model of catecholaminergic polymorphic ventricular tachycardia. EMBO Mol Med. 2012;4:180–91.
43. Kujala K, et al. Cell model of catecholaminergic polymorphic ventricular tachycardia reveals early and delayed afterdepolarizations. PLoS One. 2012;7:e44660.
44. Novak A, et al. Functional abnormalities in iPSC-derived cardiomyocytes generated from CPVT1 and CPVT2 patients carrying ryanodine or calsequestrin mutations. J Cell Mol Med. 2015;19:2006–18.
45. Preininger MK, et al. A human pluripotent stem cell model of catecholaminergic polymorphic ventricular tachycardia recapitulates patient-specific drug responses. Dis Model Mech. 2016;9:927–39.
46. Sasaki K, et al. Patient-specific human induced pluripotent stem cell model assessed with electrical pacing validates S107 as a potential therapeutic agent for Catecholaminergic polymorphic ventricular tachycardia. PLoS One. 2016;11:e0164795.
47. Zhang XH, et al. Ca2+ signaling in human induced pluripotent stem cell-derived cardiomyocytes (iPS-CM) from normal and catecholaminergic polymorphic ventricular tachycardia (CPVT)-afflicted subjects. Cell Calcium. 2013;54:57–70.
48. Lodola F, et al. Adeno-associated virus-mediated CASQ2 delivery rescues phenotypic alterations in a patient-specific model of recessive catecholaminergic polymorphic ventricular tachycardia. Cell Death Dis. 2016;7:e2393.
49. Novak A, et al. Cardiomyocytes generated from CPVTD307H patients are arrhythmogenic in response to beta-adrenergic stimulation. J Cell Mol Med. 2012;16:468–82.
50. Liang P, et al. Patient-specific and genome-edited induced pluripotent stem cell-derived cardiomyocytes elucidate single-cell phenotype of Brugada syndrome. J Am Coll Cardiol. 2016;68:2086–96.
51. Selga E, et al. Sodium channel current loss of function in induced pluripotent stem cell-derived cardiomyocytes from a Brugada syndrome patient. J Mol Cell Cardiol. 2017;114:10–9.
52. Ross SB, Fraser ST, Semsarian C. Induced pluripotent stem cell technology and inherited arrhythmia syndromes. Heart Rhythm. 2018;15(1):137–44.
53. Caballero R, et al. Flecainide increases Kir2.1 currents by interacting with cysteine 311, decreasing the polyamine-induced rectification. Proc Natl Acad Sci U S A. 2010;107:15631–6.
54. Pasca SP, et al. Using iPSC-derived neurons to uncover cellular phenotypes associated with Timothy syndrome. Nat Med. 2011;17:1657–62.
55. Tian Y, et al. Alteration in basal and depolarization induced transcriptional network in iPSC derived neurons from Timothy syndrome. Genome Med. 2014;6:75.
56. Gomez-Hurtado N, Blackwell DJ, Knollmann BC. Modelling human calmodulinopathies with induced pluripotent stem cells: progress and challenges. Cardiovasc Res. 2017;113:437–9.
57. Zhang M, et al. Recessive cardiac phenotypes in induced pluripotent stem cell models of Jervell and Lange-Nielsen syndrome: disease mechanisms and pharmacological rescue. Proc Natl Acad Sci U S A. 2014;111:E5383–92.
58. Gray B, Semsarian C, Sy RW. Brugada syndrome: a heterogeneous disease with a common ECG phenotype? J Cardiovasc Electrophysiol. 2014;25:450–6.

59. Veerman CC, et al. hiPSC-derived cardiomyocytes from Brugada syndrome patients without identified mutations do not exhibit clear cellular electrophysiological abnormalities. Sci Rep. 2016;6:30967.
60. Davis J, et al. A tension-based model distinguishes hypertrophic versus dilated cardiomyopathy. Cell. 2016;165:1147–59.
61. Broughton KM, et al. A myosin activator improves actin assembly and sarcomere function of human-induced pluripotent stem cell-derived cardiomyocytes with a troponin T point mutation. Am J Physiol Heart Circ Physiol. 2016;311:H107–17.
62. Sun N, et al. Patient-specific induced pluripotent stem cells as a model for familial dilated cardiomyopathy. Sci Transl Med. 2012;4:130ra147.
63. Wu H, et al. Epigenetic regulation of phosphodiesterases 2A and 3A underlies compromised beta-adrenergic signaling in an iPSC model of dilated cardiomyopathy. Cell Stem Cell. 2015;17:89–100.
64. Hinson JT, et al. HEART DISEASE. Titin mutations in iPS cells define sarcomere insufficiency as a cause of dilated cardiomyopathy. Science. 2015;349:982–6.
65. Karakikes I, et al. Correction of human phospholamban R14del mutation associated with cardiomyopathy using targeted nucleases and combination therapy. Nat Commun. 2015;6:6955.
66. Stillitano F, et al. Genomic correction of familial cardiomyopathy in human engineered cardiac tissues. Eur Heart J. 2016;37:3282–4.
67. Lee YK, et al. Modeling treatment response for lamin A/C related dilated cardiomyopathy in human induced pluripotent stem cells. J Am Heart Assoc. 2017;6:e005677.
68. Siu CW, et al. Modeling of lamin A/C mutation premature cardiac aging using patient-specific induced pluripotent stem cells. Aging (Albany NY). 2012;4:803–22.
69. Judge LM, et al. A BAG3 chaperone complex maintains cardiomyocyte function during proteotoxic stress. JCI Insight. 2017;2:94623.
70. Streckfuss-Bomeke K, et al. Severe DCM phenotype of patient harboring RBM20 mutation S635A can be modeled by patient-specific induced pluripotent stem cell-derived cardiomyocytes. J Mol Cell Cardiol. 2017;113:9–21.
71. Wyles SP, Hrstka SC, Reyes S, Terzic A, Olson TM, Nelson TJ. Pharmacological modulation of calcium homeostasis in familial dilated cardiomyopathy: an in vitro analysis from an RBM20 patient-derived iPSC model. Clin Transl Sci. 2016a;9:158–67.
72. Wyles SP, et al. Modeling structural and functional deficiencies of RBM20 familial dilated cardiomyopathy using human induced pluripotent stem cells. Hum Mol Genet. 2016b;25:254–65.
73. Tse HF, et al. Patient-specific induced-pluripotent stem cells-derived cardiomyocytes recapitulate the pathogenic phenotypes of dilated cardiomyopathy due to a novel DES mutation identified by whole exome sequencing. Hum Mol Genet. 2013;22:1395–403.
74. Han L, et al. Study familial hypertrophic cardiomyopathy using patient-specific induced pluripotent stem cells. Cardiovasc Res. 2014;104:258–69.
75. Lan F, et al. Abnormal calcium handling properties underlie familial hypertrophic cardiomyopathy pathology in patient-specific induced pluripotent stem cells. Cell Stem Cell. 2013;12:101–13.
76. Birket MJ, et al. Contractile defect caused by mutation in MYBPC3 revealed under conditions optimized for human PSC-cardiomyocyte function. Cell Rep. 2015;13:733–45.
77. Ojala M, et al. Mutation-specific phenotypes in hiPSC-derived Cardiomyocytes carrying either myosin-binding protein C or alpha-Tropomyosin mutation for hypertrophic cardiomyopathy. Stem Cells Int. 2016;2016:1684792.
78. Prondzynski M, et al. Evaluation of MYBPC3 trans-splicing and gene replacement as therapeutic options in human iPSC-derived cardiomyocytes. Mol Ther Nucleic Acids. 2017;7:475–86.
79. Tanaka A, et al. Endothelin-1 induces myofibrillar disarray and contractile vector variability in hypertrophic cardiomyopathy-induced pluripotent stem cell-derived cardiomyocytes. J Am Heart Assoc. 2014;3:e001263.
80. Ben Jehuda R, et al. CRISPR correction of the PRKAG2 gene mutation in a patient's induced pluripotent stem cell-derived cardiomyocytes eliminates electrophysiological and structural abnormalities. Heart Rhythm. 2018;15(2):267–76.

81. Cashman TJ, Josowitz R, Johnson BV, Gelb BD, Costa KD. Human engineered cardiac tissues created using induced pluripotent stem cells reveal functional characteristics of BRAF-mediated hypertrophic cardiomyopathy. PLoS One. 2016;11:e0146697.
82. Josowitz R, et al. Autonomous and non-autonomous defects underlie hypertrophic cardiomyopathy in BRAF-mutant hiPSC-derived cardiomyocytes. Stem Cell Reports. 2016;7:355–69.
83. Carvajal-Vergara X, et al. Patient-specific induced pluripotent stem-cell-derived models of LEOPARD syndrome. Nature. 2010;465:808–12.
84. Lee YK, et al. Modeling of Friedreich ataxia-related iron overloading cardiomyopathy using patient-specific-induced pluripotent stem cells. Pflugers Arch. 2014;466:1831–44.
85. Lee YK, et al. Efficient attenuation of Friedreich's ataxia (FRDA) cardiomyopathy by modulation of iron homeostasis-human induced pluripotent stem cell (hiPSC) as a drug screening platform for FRDA. Int J Cardiol. 2016;203:964–71.
86. Caspi O, Huber I, Gepstein A, Arbel G, Maizels L, Boulos M, Gepstein L. Modeling of arrhythmogenic right ventricular cardiomyopathy with human induced pluripotent stem cells. Circ Cardiovasc Genet. 2013;6:557–68.
87. Kim C, et al. Studying arrhythmogenic right ventricular dysplasia with patient-specific iPSCs. Nature. 2013;494:105–10.
88. Ma D, et al. Generation of patient-specific induced pluripotent stem cell-derived cardiomyocytes as a cellular model of arrhythmogenic right ventricular cardiomyopathy. Eur Heart J. 2013a;34:1122–33.
89. Lin B, et al. Modeling and study of the mechanism of dilated cardiomyopathy using induced pluripotent stem cells derived from individuals with Duchenne muscular dystrophy. Dis Model Mech. 2015;8:457–66.
90. Drawnel FM, et al. Disease modeling and phenotypic drug screening for diabetic cardiomyopathy using human induced pluripotent stem cells. Cell Rep. 2014;9:810–21.
91. Dudek J, et al. Cardiolipin deficiency affects respiratory chain function and organization in an induced pluripotent stem cell model of Barth syndrome. Stem Cell Res. 2013;11:806–19.
92. Wang G, et al. Modeling the mitochondrial cardiomyopathy of Barth syndrome with induced pluripotent stem cell and heart-on-chip technologies. Nat Med. 2014a;20:616–23.
93. Huang HP, et al. Human Pompe disease-induced pluripotent stem cells for pathogenesis modeling, drug testing and disease marker identification. Hum Mol Genet. 2011;20:4851–64.
94. Raval KK, et al. Pompe disease results in a Golgi-based glycosylation deficit in human induced pluripotent stem cell-derived cardiomyocytes. J Biol Chem. 2015;290:3121–36.
95. Sato Y, et al. Disease modeling and lentiviral gene transfer in patient-specific induced pluripotent stem cells from late-onset Pompe disease patient. Mol Ther Methods Clin Dev. 2015;2:15023.
96. Sato Y, Kobayashi H, Higuchi T, Shimada Y, Ida H, Ohashi T. Metabolomic profiling of pompe disease-induced pluripotent stem cell-derived cardiomyocytes reveals that oxidative stress is associated with cardiac and skeletal muscle pathology. Stem Cells Transl Med. 2017;6:31–9.
97. Chou SJ, et al. Energy utilization of induced pluripotent stem cell-derived cardiomyocyte in Fabry disease. Int J Cardiol. 2017;232:255–63.
98. Itier JM, et al. Effective clearance of GL-3 in a human iPSC-derived cardiomyocyte model of Fabry disease. J Inherit Metab Dis. 2014;37:1013–22.
99. Hulot JS, Jouven X, Empana JP, Frank R, Fontaine G. Natural history and risk stratification of arrhythmogenic right ventricular dysplasia/cardiomyopathy. Circulation. 2004;110:1879–84.
100. Mehta A, et al. Identification of a targeted and testable antiarrhythmic therapy for long-QT syndrome type 2 using a patient-specific cellular model. Eur Heart J. 2017;39(16):1446–55.
101. Roden DM. Cardiovascular pharmacogenomics: current status and future directions. J Hum Genet. 2016;61:79–85.
102. Roden DM. Long QT syndrome: reduced repolarization reserve and the genetic link. J Intern Med. 2006;259:59–69.

103. Stillitano F, et al. Modeling susceptibility to drug-induced long QT with a panel of subject-specific induced pluripotent stem cells. elife. 2017;6:e19406.
104. Burridge PW, et al. Human induced pluripotent stem cell-derived cardiomyocytes recapitulate the predilection of breast cancer patients to doxorubicin-induced cardiotoxicity. Nat Med. 2016;22:547–56.
105. Bedada FB, Wheelwright M, Metzger JM. Maturation status of sarcomere structure and function in human iPSC-derived cardiac myocytes. Biochim Biophys Acta. 2016;1863:1829–38.
106. Eder A, Vollert I, Hansen A, Eschenhagen T. Human engineered heart tissue as a model system for drug testing. Adv Drug Deliv Rev. 2016;96:214–24.
107. Tiburcy M, et al. Defined engineered human myocardium with advanced maturation for applications in heart failure modeling and repair. Circulation. 2017;135:1832–47.
108. Turnbull IC, et al. Advancing functional engineered cardiac tissues toward a preclinical model of human myocardium. FASEB J. 2014;28:644–54.
109. Lu J, et al. An abnormal TRPV4-related cytosolic Ca2+ rise in response to uniaxial stretch in induced pluripotent stem cells-derived cardiomyocytes from dilated cardiomyopathy patients. Biochim Biophys Acta. 2017;1863:2964–72.

Part III
Molecular Technologies

Chapter 6
Endogenous Signal-Responsive Transgene Switch Systems for Visualization and Purification of Specific Cells

Hideyuki Nakanishi and Hirohide Saito

Abstract Induced pluripotent stem cells (iPSCs) are induced to differentiate into target cells, and these differentiated cells are used for regenerative medicines or drug discovery. Using endogenous signal-responsive transgene switch systems, scientists can visualize different differentiation states and purify targeted cell types. These systems are composed of transgenes and binding domains of cell-specific proteins or miRNAs to detect endogenous molecules and regulate transgene expression. In this chapter, we describe the designs, operations, and applications of switch systems.

Keywords RNA binding protein · microRNA · Synthetic biology · Stem cell Differentiation · Imaging · Cell purification · Reporter gene · In vitro transcription Regenerative medicine · Modified mRNA · Translation

6.1 Introduction

iPSCs have the potential to differentiate into various cell types. These differentiated cells can be used for many applications including regenerative medicine and drug discovery. However, because the differentiation process is imperfect, unwanted cells such as undifferentiated or differently differentiated cells can contaminate the population and cause adverse effects such as tumor formations, loss of therapeutic effects, or erroneous drug evaluation.

To overcome this problem, our group has recently developed transgene-expression switch systems that are composed of engineered messenger RNAs (mRNAs) or plasmid DNAs (pDNAs). These nucleic acids respond to cell state-specific signals, such

H. Nakanishi · H. Saito (✉)
Department of Life Science Frontiers, Center for iPS Cell Research and Application (CiRA), Kyoto University, Kyoto, Japan
e-mail: hsaito-g@cira.kyoto-u.ac.jp

H. Inoue, Y. Nakamura (eds.), *Medical Applications of iPS Cells*, Current Human Cell Research and Applications,
https://doi.org/10.1007/978-981-13-3672-0_6

as endogenous proteins or microRNAs (miRNAs). Thus, these systems enable us to visualize the cell state during differentiation and purify specific cell populations, such as iPSC-derived cardiomyocytes or neurons. In this chapter, we describe the design and application of both miRNA- and protein-responsive transgene switch systems.

6.2 Visualization of Differentiation States and Purification of Specific Cells Using Fluorescent Protein Expression Switches

6.2.1 Endogenous Protein-Responsive Switches that Regulate Fluorescent Protein Expression

Various cell-specific RNA binding proteins regulate translation, miRNA biogenesis, mRNA localization, and other biomolecular phenomena that maintain or alter the cellular state [1–4]. Our group previously developed a transgene switch system that responds to the archaeal RNA binding protein L7Ae [5–10]. In this system, pDNAs or in vitro transcribed mRNAs that contain an L7Ae-binding motif, kink-turn (K-turn), in their 5′ untranslated regions (UTRs) are transfected into cells. L7Ae is expressed by the transfected cells and binds to the K-turn to inhibit mRNA translation. Although this system is effective at repressing translation, visualization of the differentiation state of mammalian cells is difficult because L7Ae is not a mammalian protein.

Therefore, our group has also developed a mRNA switch that responds to human endogenous LIN28A [11], which is a highly expressed RNA binding protein in human pluripotent stem cells (PSCs). The original LIN28A-binding RNA sequence (motif) derived from preE-let-7d miRNA was inserted into the 5′ UTR of mRNA that coded a fluorescent protein, Azami-Green. The modified Azami-Green mRNA was co-transfected with reference mRNA coding another fluorescent protein, Kusabira Orange. We expected that any interaction between LIN28A and the LIN28A-binding RNA motif derived from preE-let-7d would repress translation of the Azami-Green-encoding mRNA and decrease the Azami-Green/Kusabira Orange expression ratio in Lin28A-expressing cells, making it possible to distinguish Lin28A-positive and Lin28A-negative cells by flow cytometry (Fig. 6.1a). However, the original Lin28A-binding RNA motif did not efficiently repress translation of the Azami-Green-encoding mRNA in Lin28A-expressing cells. As a next step, we stabilized the secondary structure of the LIN28A-binding motif by increasing the GC-content of the stem region. The new Lin28A-binding motif (stbC) efficiently repressed translation of the Azami-Green mRNA and clearly distinguished two types of cells, human iPSCs (hiPSCs) and differentiated cells (Fig. 6.1b).

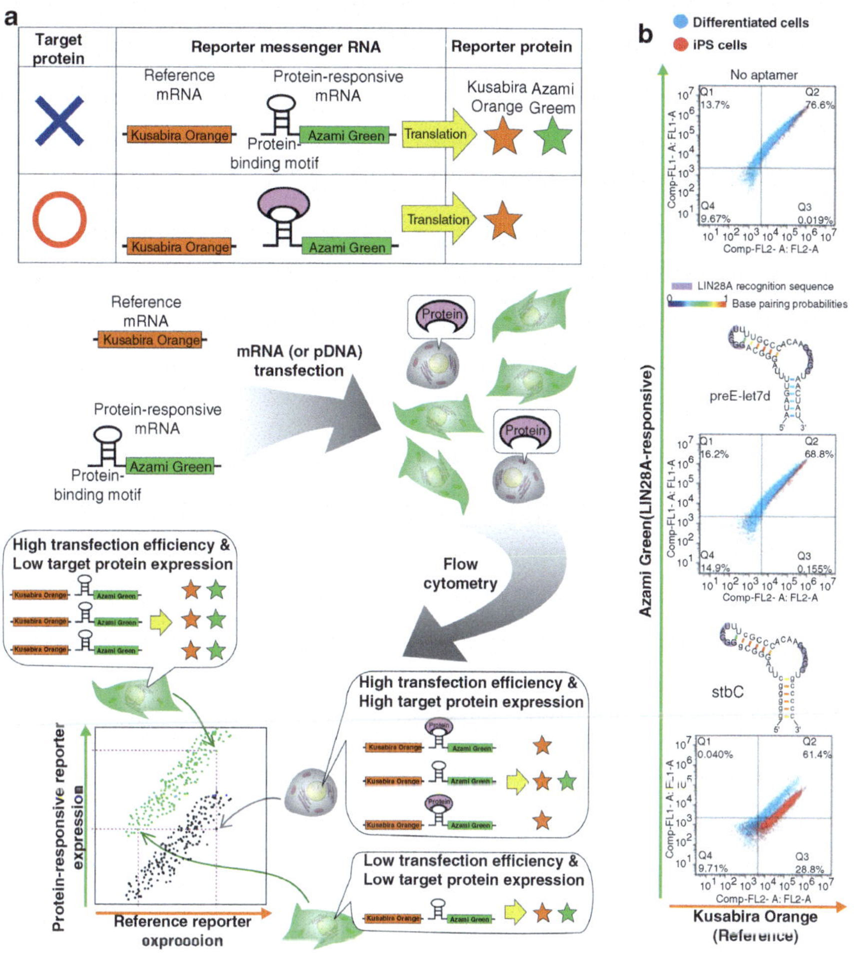

Fig. 6.1 Distinction of iPSCs and differentiated cells using LIN28A-responsive switches. (**a**) Scheme to distinguish cells using protein-responsive switches. The translation of protein-responsive mRNAs is repressed by the binding of target proteins at protein binding motifs. Reference mRNAs are co-transfected to compensate transfection efficiencies. (**b**) The distinction of iPSCs and differentiated cells by using original (preE-let7d) or stabilized (stbC) LIN28A-binding motifs

6.2.2 *miRNA-Responsive Fluorescent Protein Expression Switch*

miRNAs are small noncoding RNAs that repress translation or induce mRNA degradation by binding to complementary (or partially complementary) mRNA sequences. To date, more than 2500 miRNAs have been reported in human cells, and their activity differs among cell types [12].

We have developed miRNA-responsive, fluorescent protein expression switches to distinguish cell types [13–16]. To construct miRNA-responsive switches, we inserted the complete complementary sequences of miRNAs into the 5′ or 3′ UTRs of mRNAs that code fluorescent proteins. If switch-transfected cells highly express the target miRNAs, these miRNAs should bind their complementary sequences and inhibit translation of the fluorescent protein by cleaving the mRNA.

As in the case of protein-responsive switches, miRNA-responsive mRNAs were co-transfected with mRNA coding another fluorescent protein (reference mRNA), and the expression ratio of the two fluorescent proteins was analyzed by flow cytometry. To distinguish and purify cardiomyocytes derived from hiPSCs, as an example, mRNAs with complementary sequences for hsa-miR-1-3p, 208a-3p, and 499a-5p, which are highly active in human cardiomyocytes, were synthesized by in vitro transcription [13]. miRNA-responsive mRNAs coding blue fluorescent protein (BFP) were co-transfected with reference EGFP mRNA into hiPSC-derived cells including cardiomyocytes and other cell types. Flow cytometry analysis revealed two distinct cell populations based on the BFP/EGFP expression ratio. Cells were sorted by fluorescence-activated cell sorting (FACS) according to the BFP/EGFP ratio (Fig. 6.2), and the expression of cardiac troponin T (cTNT), which is a marker protein of cardiomyocytes, in each population was analyzed. While cTNT-positive cells were abundant in the low BFP/EGFP ratio (i.e., high miRNA activity) population, these cells were rare in the high BFP/EGFP ratio (i.e., low miRNA activity) population. Therefore, miRNA-responsive switches enable not only the distinction of cell types but also the sorting of types. The effectiveness of miRNA-responsive switch-mediated sorting was also validated by the sorting of endothelial cells, hepatocytes, and insulin-producing cells when using switches that respond to miRNAs specific to each cell type [13].

miRNA-responsive switches also enable the continuous visualization of the differentiation process. For this purpose, transfection of in vitro transcribed mRNAs is not suitable because of their short expression time. Therefore, we constructed DNA-based integrative vectors that express both miRNA-responsive Azami-Green and reference tagRFP mRNAs (Fig. 6.3a) [16]. To transcribe the same amount of tagRFP and Azami-Green mRNAs, the vector is designed to transcribe the two reporter genes as a single mRNA. After transcription, the two reporter genes are separated by RNaseP/Z-mediated cleavage. For efficient genomic integration, the reporter genes were inserted between two inverted repeat sequences (IRs) of the *piggyBac* transposon. When the vector is co-transfected with *piggyBac* transposase expression vector, the transposase excises the region enclosed by the two IRs from a pDNA and integrates it into a genome. For continuous visualization of hiPSC differentiation, we constructed a vector that responds to hsa-miR-302a-5p, which is highly active in PSCs but not in differentiated cells. After the transfection and integration of the vector, spontaneous differentiation of hiPSCs was induced, and the fluorescent microscopic images of these cells were continuously captured. Because hsa-

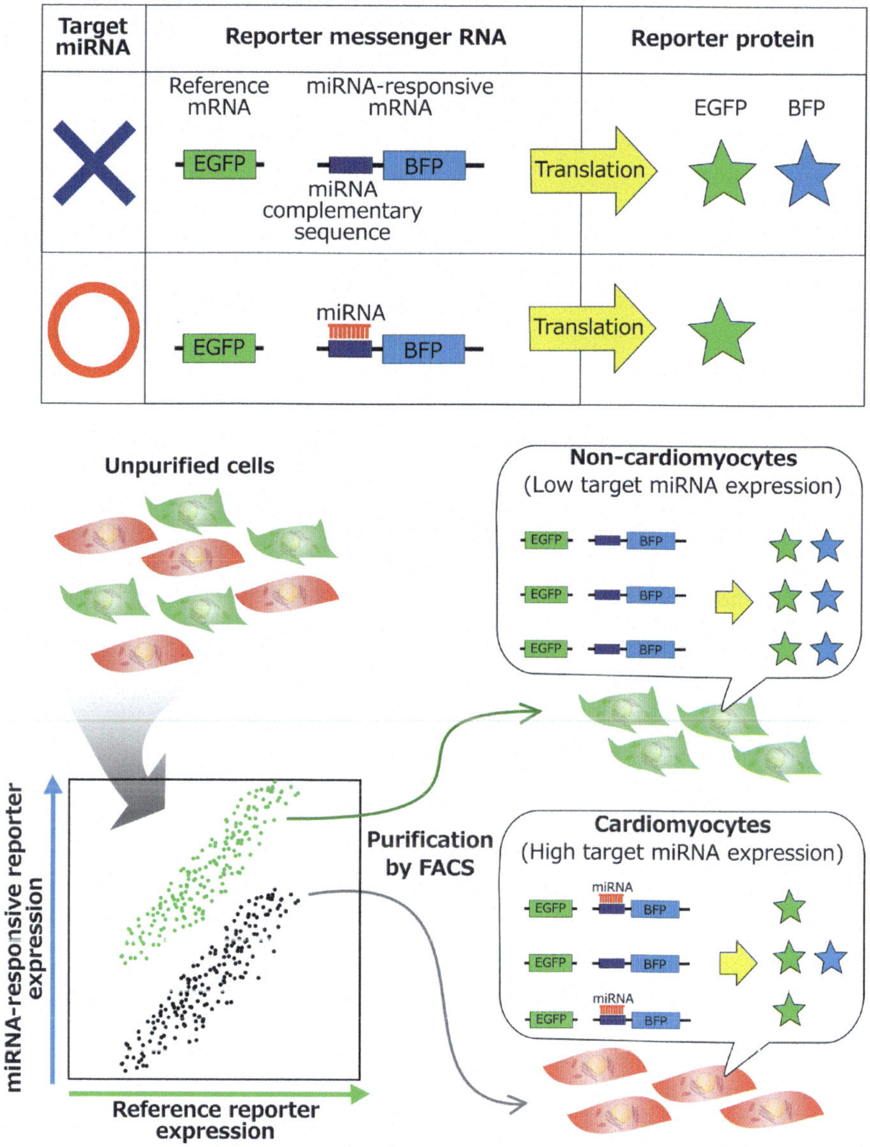

Fig. 6.2 FACS-mediated purification of cardiomyocytes using miRNA-responsive switches. The translation of miRNA-responsive mRNAs is repressed by the binding of target miRNAs at their complementary sequences. Reference mRNAs are co-transfected to compensate transfection efficiencies. hiPSC-derived cells including cardiomyocytes were transfected with cardiomyocyte-specific miRNA-responsive switches and sorted based on the fluorescent protein expression

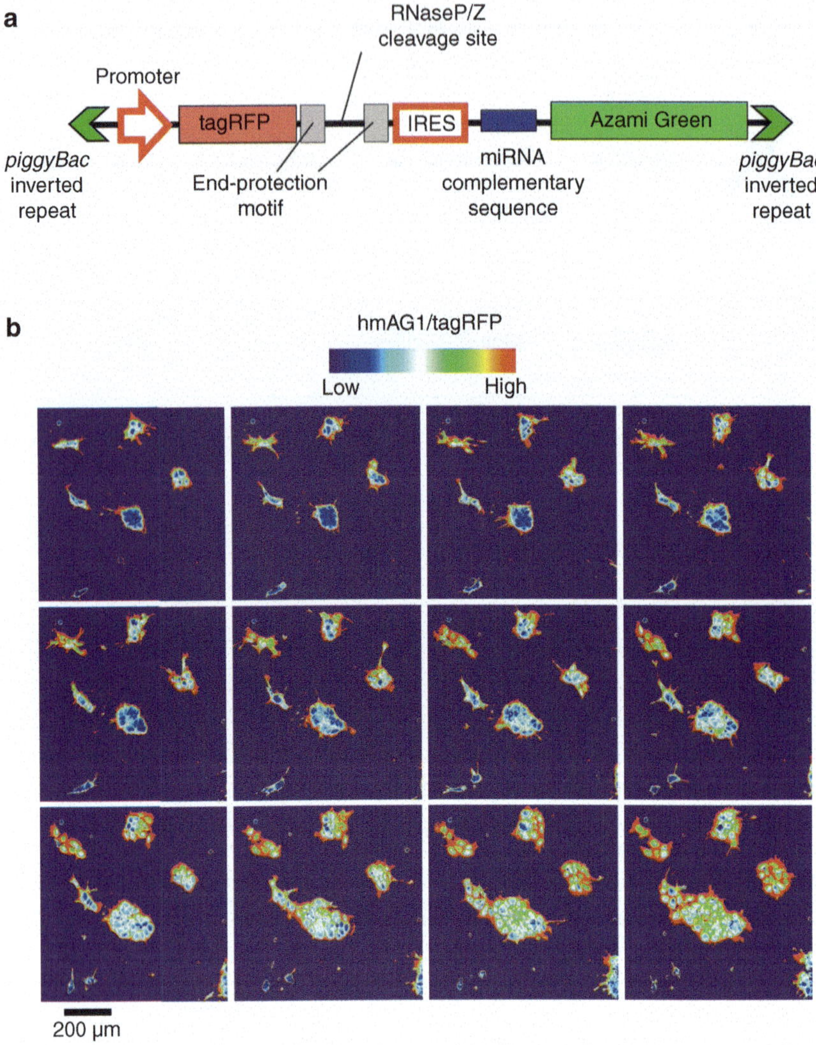

Fig. 6.3 Continuous visualization of differentiation using a *piggyBac*-based miRNA-responsive switch. (**a**) The vector design of *piggyBac*-based miRNA-responsive switches. Reference and miRNA-responsive reporters are transcribed as a single mRNA. After transcription, RNaseP/Z-mediated cleavage separates the two reporter genes. (**b**) Time-lapse images of differentiating cells. Differentiation was induced by culturing the cells in medium lacking basic FGF. Images were captured every 3 h, and the ratio of hmAG1/tagRFP is shown as pseudo-colors

miR-302a-5p activity decreases during differentiation, the miRNA-mediated repression of Azami-Green is alleviated, and the Azami-Green/tagRFP ratio is increased. Calculation of the ratio enabled the visualization of the differentiation time course of hiPSCs (Fig. 6.3b). The vector can also be adapted for the visualization of differentiation to specific cells by simply exchanging the complementary sequence of the miRNA. For example, we visualized cardiomyocytes using a hsa-miR-208a-3p-responsive switch vector [16]. Because the DNA-based integrative switch system enables repetitive checks of the differentiation status by a single transfection, it reduces the labor and cost of nonclinical applications such as screenings for optimal differentiation conditions.

6.3 Elimination of Undesired Cells Using Cell-Specific Killing Switches

6.3.1 Elimination of miRNA-Positive Cells Using an Antibiotic Gene Expression Switch

Because FACS is unsuitable for the large-scale purification of cells, our group has developed cell-selective killing switches that enable the purification of specific cells as an alternative. The contamination of undifferentiated iPSCs risks causing tumor formation after cell transplantation. For the selective killing of iPSCs, we designed miRNA-responsive puromycin-resistant gene expression switch. The switch mRNA codes the puromycin-resistant gene and contains the complementary sequence for hsa-miR-302a-5p at the 5′ UTR. Because hsa-miR-302a-5p is highly active in iPSCs, the translation of the puromycin-resistant gene is repressed to leave the cells puromycin-sensitive. In contrast, the activity of hsa-miR-302a-5p is low in differentiated cells. Therefore, the puromycin-resistant gene is efficiently translated, which makes the cells puromycin-resistant. To selectively eliminate iPSCs, we added the switch to the coculture of iPSC-derived neurons and undifferentiated iPSCs. Transfection of the switch and subsequent treatment with puromycin selectively eliminated TRA-1-60-positive iPSCs (Fig. 6.4a) [15].

6.3.2 Elimination of miRNA-Negative Cells Using Apoptosis-Inducing Gene Expression Switches

A system that eliminates miRNA-negative cells is useful for purifying desired miRNA-positive cells. Therefore, we also designed a switch that selectively eliminates target miRNA-negative cell populations. The switch system is composed of puromycin-resistant gene-coding mRNA and pro-apoptotic protein (Bim)-coding mRNA. While Bim mRNA contained miRNA target sites, puromycin-resistant gene-coding mRNA did not.

To purify iPSC-derived cardiomyocytes, we prepared Bim-coding switch mRNAs that contain target sites for hsa-miR-1-3p or 208a-3p, which are highly active in human cardiomyocytes. While hsa-miR-1-3p- or 208a-3p-negative cells were killed by the Bim-mediated induction of apoptosis, cardiomyocytes (hsa-miR-1-3p- and

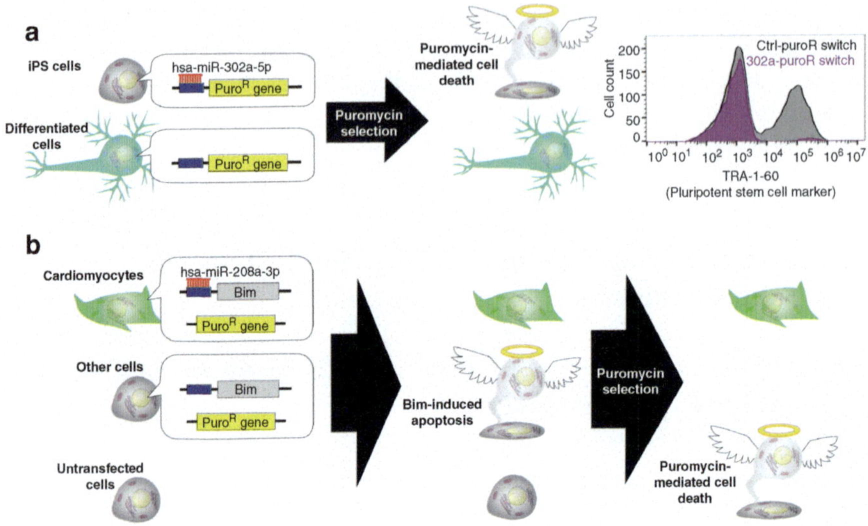

Fig. 6.4 Selective elimination of miRNA-positive or miRNA-negative cells using miRNA-responsive switches. (**a**) Selective elimination of miRNA-positive cells by an antibiotic-resistant gene-coding switch. iPSCs and differentiated cells were transfected with hsa-miR-302a-5p-responsive switch coding the puromycin-resistant gene. Puromycin-resistant gene is selectively repressed in iPSCs in which hsa-miR-302a-5p is highly active. Immunostaining of TRA-1-60 (a marker of PSCs) showed selective elimination of iPSCs after treatment with the switch and puromycin. The histogram was modified from Parr et al. [15] under a Creative Commons Attribution 4.0 International License (http://creativecommons.org/licenses/by/4.0/). (**b**) Selective elimination of miRNA-negative cells by an apoptosis-inducing gene-coding switch. iPSC-derived cells including cardiomyocytes were transfected with the puromycin-resistant gene-coding mRNA and hsa-miR-208a-3p-responsive switch coding Bim gene. In cardiomyocytes, Bim expression is selectively repressed by hsa-miR-208a-3p

miR-208a-3p-positive) survived because of the miRNA-mediated repression of Bim. In addition, it is important to eliminate RNA-untransfected cells. We therefore co-transfected puromycin-resistant gene-coding mRNA together with Bim switch mRNA. Although apoptosis is not induced in untransfected miRNA-negative cells that did not express both Bim and puromycin-resistant genes, these cells are eliminated by the addition of puromycin (Fig. 6.4b). Therefore, we could successfully enrich cardiomyocytes without using FACS by the transfection of Bim and puromycin-resistant mRNAs and the addition of puromycin [13].

6.4 Cell-Selective Genome Editing

Genome editing is a precise method to modify genomes. In genome editing, the genome is cleaved by sequence-specific DNA cutting enzymes such as zinc finger nucleases (ZFNs), transcription activator-like effector nucleases (TALENs), or

Cas9. Among these enzymes, Cas9 and Cas9-related proteins are currently attracting the greatest attention because of their easiness to adapt arbitrary sequences. While ZFNs and TALENs recognize their target sequences by the proteins themselves, Cas9 makes a complex with guide RNA (gRNA), and the target-sequence recognition depends on the gRNA. Therefore, we can make Cas9 cleave arbitrary target sequences by simply exchanging a short (around 20 bp) target sequence of gRNA [17]. After the genome cleavage, cellular machinery repairs the genomes by nonhomologous end joining (NHEJ) or homologous recombination (HR). During the repair, sequences can be inserted into or deleted from the genome. Therefore, various alterations of the genome, including the knock-in of transgenes and knockout of endogenous genes, can be achieved.

Combining genome editing technology with our miRNA-responsive switch, we have recently developed a cell-selective genome editing system [18]. In this system, the target sites of miRNAs were added to the mRNA of Cas9. As was the case for fluorescent proteins or Bim, Cas9 expression is repressed in miRNA-positive cells. Therefore, Cas9-mediated genome editing selectively occurs in miRNA-negative cells. We designed Cas9 mRNA with the target sequence of hsa-miR-302a-5p and co-transfected it into EGFP-expressing HeLa and hiPSCs with an EGFP-targeting gRNA. As mentioned in Sect. 6.2, hsa-miR-302a-5p is highly active in hiPSCs. Therefore, Cas9-mediated knockout of the EGFP gene was repressed in iPSCs, but efficient knockout of the gene was achieved in HeLa cells (Fig. 6.5).

This cell-selective genome editing system will enable various applications in basic studies and clinical therapies. For example, undesired cells can be killed by the knockout of housekeeping genes or knock-in of apoptosis-inducing genes. The knockout of embryonic lethal genes in specific cells in transgenic animals and the avoidance of undesired genome editing in germ cells in gene therapy are also promising applications.

6.5 Design and Preparation of Suitable Switches

Transgene switches can be transfected as either in vitro transcribed mRNAs or DNA vectors. Both vectors have their advantages and limitations, and the most suitable vector depends on the application.

In vitro transcribed mRNAs are safer than DNA-based vectors because of short-term expression time and the absence of genomic integration. Therefore, they are suitable for clinical application for which safety is paramount. mRNA transfection is also appropriate to minimize side effects of the transgenes on the cellular state. To prepare mRNA-based switches, first, DNA templates are designed. The DNA templates should contain T7 promoter sequence, 5′ UTR, a coding region of a suitable gene, 3′ UTR, and polyA tail (not polyadenylation signals) sequences. Using the DNA template and T7 polymerase, mRNAs are prepared by in vitro transcription. Modified nucleotides such as pseudouridine and 5-methylcytosine can be incorporated into the mRNAs to avoid immune responses. 5′ cap analogs such as Anti-Reverse

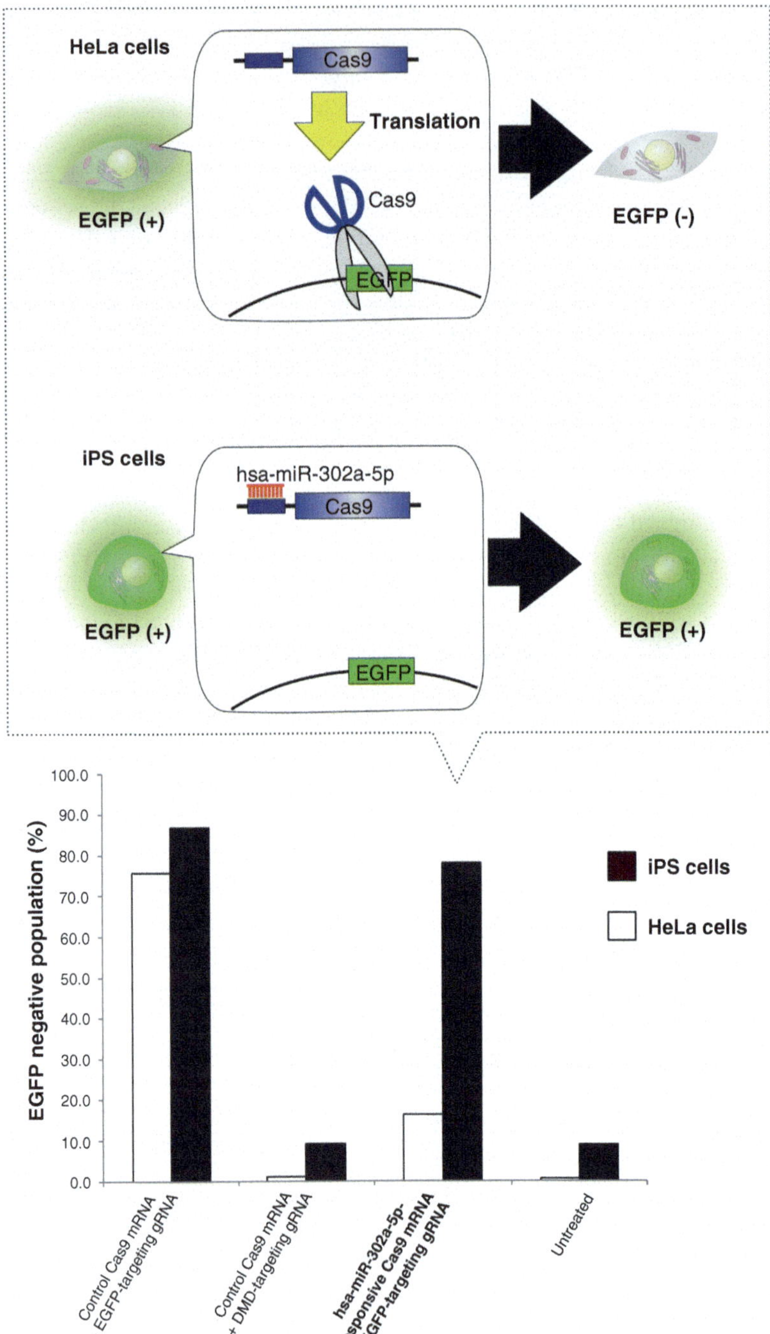

Fig. 6.5 Cell-selective genome editing by Cas9-coding switch. HeLa cells and iPSCs were transfected with hsa-miR-302a-5p-responsive switch coding Cas9. In iPSCs, Cas9 expression is repressed by hsa-miR-302a-5p. Eventually, knockout of EGFP by Cas9 was selectively performed in HeLa cells

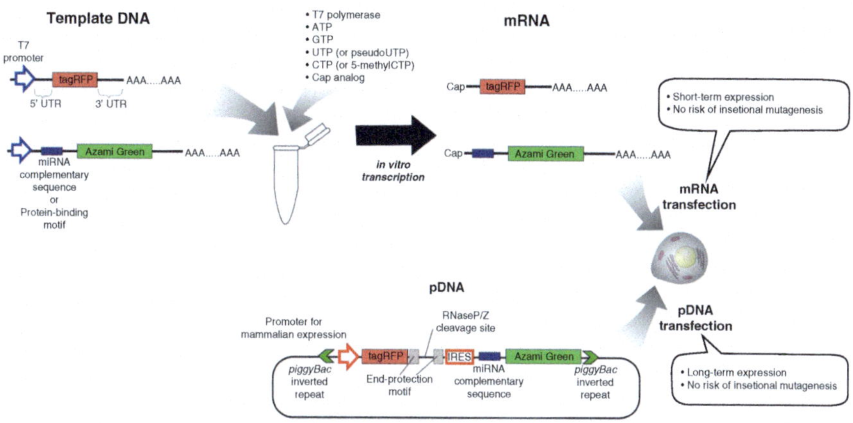

Fig. 6.6 Preparation of mRNA- and DNA-based switches. The appropriate switch depends on the application. While mRNA-based switches are relatively safer, DNA-based switches enable continuous transgene expression. In a preparation of mRNA-based switches, mRNAs are transcribed in vitro from DNA templates containing a T7 promoter, a coding region, 5′ and 3′ UTRs, and polyA tail

Cap Analog (ARCA) are also added to the reaction mixture. After in vitro transcription, DNA templates are digested by DNase, and mRNAs are dephosphorylated by alkaline phosphatase. The length of the obtained mRNAs can be confirmed by electrophoresis. After the preparation, the mRNAs can be transfected into the cells by transfection reagents such as Stemfect RNA Transfection Kit (Stemgent) or Lipofectamine MessengerMAX (Thermo Fisher Scientific) (Fig. 6.6).

DNA-based vectors risk insertional mutagenesis by the genomic integration of transgenes. However, they enable long-term expression of the transgenes, especially when transposon-based or episomal vectors are used. Therefore, they are suitable for continuous visualization of the differentiation process. Different from direct mRNA transfection, the expression from DNA-based vectors can be affected at the transcriptional level. Especially for long-term continuous visualization, epigenetic modifications such as DNA methylation or histone modifications can change the expression ratio of the miRNA-responsive and reference reporter genes, making it difficult to visualize actual miRNA activity. To avoid this problem, we designed a vector that transcribes these two reporter genes as a single mRNA, followed by RNaseP/Z-mediated separation of the two genes [16] (Figs. 6.3 and 6.6). The selection of promoters is also important. While CMV and RSV promoters are conventionally used in mammalian cells, they are not suitable for use in iPSCs [19]. To achieve efficient expression in iPSCs, CAG or EF1 promoters are recommended.

To enhance the sensitivity of miRNA-responsive switches, several strategies can be adopted. First, the target sites of the miRNAs can be inserted into either 5′ or 3′ UTR or both. Insertions into 5′ UTRs make the switch more sensitive than into 3′ UTRs. Second, insertions of multiple miRNA target sites can enhance the sensitivity of switches (Fig. 6.7a). RNA secondary structures may also affect the sensitivity for target miRNA detection, but the design of highly sensitive switches based on the regulation of the RNA structure remains difficult [16].

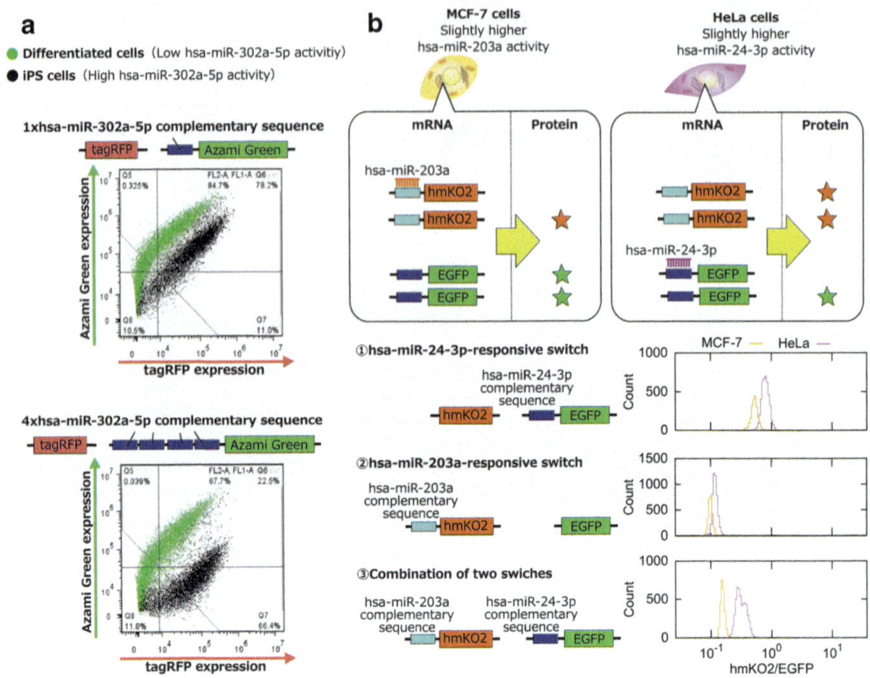

Fig. 6.7 Improvements of miRNA-responsive switches. (**a**) miRNA-responsive switch performance can be improved by increasing the copy number of miRNA-complementary sequences. (**b**) Better resolution of different cell populations can be achieved by the co-transfection of multiple switches. The histogram was modified from Endo et al. [14] under a Creative Commons Attribution 4.0 International License (http://creativecommons.org/licenses/by/4.0/)

If a single miRNA is not sufficient to distinguish the target cells from other cells, co-transfection of multiple switches is a possible solution. For example, when comparing HeLa and MCF-7 cells, only small differences in hsa-miR-24-3p and hsa-miR-203a activity between the two types of cells were observed. hsa-miR-24-3p activity was 1.3-fold higher in HeLa cells than in MCF-7 cells, and hsa-miR-203a activity was 1.5-fold higher in MCF-7 cells than in HeLa cells. The miRNA activities of each were insufficient to distinguish these two cell types when the two switches were used individually. However, co-transfection of both switches and calculation of the ratio of the miRNA activities enabled clear resolution [14] (Fig. 6.7b).

6.6 Conclusion and Future Perspective

The endogenous signal-responsive switch system has several advantages over conventional visualization or purification tools such as genomic transgene knock-in or antibodies.

Compared with transgene knock-in, the switch system is relatively safe and easy to use. When transfecting the switch system as in vitro transcribed mRNAs, the mRNAs cannot be integrated into the genome. Therefore, different from transgene knock-in, the switch system can be used without the risk of genomic alterations. This feature is especially advantageous when visualizing or purifying cells for clinical use such as cell transplantation. While fluorescent dye-fused antibodies can also be used without genomic alterations, the switch system has an advantage in flexibility. It enables not only the visualization and purification based on fluorescence, but also the elimination of harmful or unnecessary cells without using FACS. We have previously adopted fluorescent protein, antibiotic-resistant, apoptosis-inducing, and Cas9 genes for the switch system. Other suitable genes can also be used by simply exchanging the coding region of the mRNA or pDNA in the switch system.

In addition, the switch system is easy to design. miRNA-responsive switches can be adapted to any cell-specific miRNAs by just inserting the complementary sequences of the miRNAs. The design of protein-responsive switches is not as easy as miRNA-responsive switches, but the screening of protein-binding RNA motifs by systematic evolution of ligands by exponential enrichment (SELEX) [20, 21] and the stabilization of the secondary structures of these motifs based on RNA secondary structure predictions [22, 23] may help to design them.

These advantageous features of the switch system will enable various applications regarding iPSCs. For example, apoptosis-inducing or drug-resistant gene-coding switches enable the efficient purification of differentiated cells for transplantation or drug screenings. Visualization of the differentiation efficiency by fluorescent protein-coding switches will help screen efficient differentiation protocols and elucidate differentiation mechanisms.

Acknowledgment We are grateful to Dr. Shunsuke Kawasaki and Mr. Moe Hirosawa for assisting figure preparations. We also thank Dr. Peter Karagiannis (Kyoto University) for English proofreading.

References

1. Chen Q, Hu G. Post-transcriptional regulation of the pluripotent state. Curr Opin Genet Dev. 2017;46:15–23.
2. Hao J, Duan FF, Wang Y. MicroRNAs and RNA binding protein regulators of microRNAs in the control of pluripotency and reprogramming. Curr Opin Genet Dev. 2017;46:95–103.
3. Medioni C, Mowry K, Besse F. Principles and roles of mRNA localization in animal development. Development. 2012;139:3263–76.
4. Martin KC, Ephrussi A. mRNA localization: gene expression in the spatial dimension. Cell. 2009;136:719–30.
5. Saito H, Kobayashi T, Hara T, Fujita Y, Hayashi K, Furushima R, Inoue T. Synthetic translational regulation by an L7Ae-kink-turn RNP switch. Nat Chem Biol. 2010;6:71–8.
6. Saito H, Fujita Y, Kashida S, Hayashi K, Inoue T. Synthetic human cell fate regulation by protein-driven RNA switches. Nat Commun. 2011;2:160.

7. Stapleton JA, Endo K, Fujita Y, Hayashi K, Takinoue M, Saito H, Inoue T. Feedback control of protein expression in mammalian cells by tunable synthetic translational inhibition. ACS Synth Biol. 2012;1:83–8.
8. Endo K, Stapleton JA, Hayashi K, Saito H, Inoue T. Quantitative and simultaneous translational control of distinct mammalian mRNAs. Nucleic Acids Res. 2013;41:1–12.
9. Endo K, Hayashi K, Inoue T, Saito H. A versatile cis-acting inverter module for synthetic translational switches. Nat Commun. 2013;4:2393.
10. Wroblewska L, Kitada T, Endo K, Siciliano V, Stillo B, Saito H, Weiss R. Mammalian synthetic circuits with RNA binding proteins for RNA-only delivery. Nat Biotechnol. 2015;33:839–41.
11. Kawasaki S, Fujita Y, Nagaike T, Tomita K, Saito H. Synthetic mRNA devices that detect endogenous proteins and distinguish mammalian cells. Nucleic Acids Res. 2017;45:1–13.
12. Kozomara A, Griffiths-Jones S. MiRBase: integrating microRNA annotation and deep-sequencing data. Nucleic Acids Res. 2011;39:D152–7.
13. Miki K, Endo K, Takahashi S, et al. Efficient detection and purification of cell populations using synthetic microRNA switches. Cell Stem Cell. 2015;16:699–711.
14. Endo K, Hayashi K, Saito H. High-resolution identification and separation of living cell types by multiple microRNA-responsive synthetic mRNAs. Sci Rep. 2016;6:21991.
15. Parr CJC, Katayama S, Miki K, Kuang Y, Yoshida Y, Morizane A, Takahashi J, Yamanaka S, Saito H. microRNA-302 switch to identify and eliminate undifferentiated human pluripotent stem cells. Sci Rep. 2016;6:32532.
16. Nakanishi H, Miki K, Komatsu KR, Umeda M, Mochizuki M, Inagaki A, Yoshida Y, Saito H. Monitoring and visualizing microRNA dynamics during live cell differentiation using microRNA-responsive non-viral reporter vectors. Biomaterials. 2017;128:121–35.
17. Jinek M, Chylinski K, Fonfara I, Hauer M, Doudna JA, Charpentier E. A programmable dual-RNA–guided DNA endonuclease in adaptive bacterial immunity. Science. 2012;337:816–22.
18. Hirosawa M, Fujita Y, Parr CJC, Hayashi K, Kashida S, Hotta A, Woltjen K, Saito H. Cell-type-specific genome editing with a microRNA-responsive CRISPR–Cas9 switch. Nucleic Acids Res. 2017;45:e118.
19. Tashiro K, Inamura M, Kawabata K, Sakurai F, Yamanishi K, Hayakawa T, Mizuguchi H. Efficient adipocyte and osteoblast differentiation from mouse induced pluripotent stem cells by adenoviral transduction. Stem Cells. 2009;27:1802–11.
20. Tuerk C, Gold L. Systematic evolution of ligands by exponential enrichment: RNA ligands to bacteriophage T4 DNA polymerase. Science. 1990;249:505–10.
21. Darmostuk M, Rimpelova S, Gbelcova H, Ruml T. Current approaches in SELEX: an update to aptamer selection technology. Biotechnol Adv. 2014;33:1141–61.
22. Kato Y, Sato K, Asai K, Akutsu T. Rtips: fast and accurate tools for RNA 2D structure prediction using integer programming. Nucleic Acids Res. 2012;40:29–34.
23. Kawaguchi R, Kiryu H. Parallel computation of genome-scale RNA secondary structure to detect structural constraints on human genome. BMC Bioinformatics. 2016;17:203.

Chapter 7
Precision Genome Editing in Human-Induced Pluripotent Stem Cells

Knut Woltjen

Abstract The advent of human-induced pluripotent stem cell (hiPSC) technology enabled researchers to gain access to the various somatic cell types that comprise the human body, along with the underlying genetic code which details their function or dysfunction in normal and disease states, respectively. In vitro disease models based on patient-specific hiPSCs have already demonstrated great potential for elucidating disease mechanisms, drug discovery, and validation of cell replacement therapies. In certain cases, accurate recapitulation or rectification of the genetic causes of disease requires the ability to precisely modify just a single base of DNA among the billions present in the human genome. Together with recent advances in gene editing technologies such as programmable endonucleases, we are now able to re-create pathogenic mutations with base precision for more reliable disease models. Moreover, these combined methods have opened the door to scarless repair of disease alleles for the future of personal stem cell therapy.

Keywords Human-induced pluripotent stem cell · hiPSC · Gene targeting · Nuclease · CRISPR/Cas9 · HDR · NHEJ · MMEJ · Disease model · Gene correction · Isogenic control

7.1 Introduction

Human-induced pluripotent stem cells (hiPSCs) share fundamental properties with human embryonic stem cells (hESCs), as well as provide new opportunities in biological and medical applications [1]. Key similarities include characteristics such as indefinite replication, which addresses the availability of cells as a resource; clonogenicity, which has significant implications for safety through cell screening,

K. Woltjen (✉)
Center for iPS Cell Research and Application (CiRA), Kyoto University, Kyoto, Japan

Hakubi Center for Advanced Research, Kyoto University, Kyoto, Japan
e-mail: woltjen@cira.kyoto-u.ac.jp

H. Inoue, Y. Nakamura (eds.), *Medical Applications of iPS Cells*,
Current Human Cell Research and Applications,
https://doi.org/10.1007/978-981-13-3672-0_7

selection, and amplification; and stable maintenance of differentiation potential, enabling the human to be used as an in vitro genetic model via tissue surrogates and organoids [2]. Uniquely, hiPSCs donated by healthy or diseased individuals simultaneously circumvented many ethical debates surrounding hESCs and enabled the development and banking of personal and public stem cell resources for prospective therapeutics and disease modeling [3]. While the underlying complexities of human genetic variation brought the need for isogenic controls in genotype-phenotype correlations, the rise of modern genome editing tools has helped address this need.

Recent advances in the development of programmable endonucleases such as zinc-finger nucleases (ZFNs), transcription activator-like effector nucleases (TALENs), and clustered regularly interspaced short palindromic repeats (CRISPR)/Cas9 [4] have enabled researchers to more easily manipulate the genome of healthy donor or patient-derived hiPSCs [5, 6]. With such tools in hand, we are free to create mutations associated with disease pathogenesis for in vitro modeling or even correct causative mutations for eventual autologous therapeutic applications using a patient's own hiPSCs or differentiated derivatives. This chapter will provide an overview of programmable nuclease applications to stimulate cellular DNA repair for the purpose of precisely modifying target genomic loci. How nuclease delivery methods and the identity of the designer DNA template impact the process will be discussed. Methods designed to enrich desirable mutations will be highlighted, along with some of the peculiarities of gene editing using the hiPSC as a genetic model.

7.2 Creating Genomic DSBs

It has long been known that the induction of a double-strand break (DSB) using site-specific restriction endonucleases significantly stimulates endogenous DNA repair responses at transgenic target sites [7]. Subsequent efforts to generate recombinant proteins capable of targeting endogenous genomic sites employed a combined knowledge of zinc-finger DNA binding proteins [8] and the separable DNA cleavage domain from *Fok*I [9], giving rise to the first programmable nucleases, the ZFNs (Fig. 7.1a) [10]. The early days of ZFNs were faced with challenges in zinc finger assembly and difficulties in de novo prediction of DNA binding parameters due to the complexities of neighboring finger: finger interactions [11], such that assisted gene targeting in hESCs and hiPSCs with ZFNs, remained a method reserved primarily to the elite [12]. In 2009, the 1:1 protein/DNA binding code for transcription activator-like effectors (TALEs) from plant pathogenic bacteria gave rise to a new species of simpler programmable DNA binding domains (Fig. 7.1a) [13, 14]. Combined with the same dimerizing *Fok*I nuclease domain employed by ZFNs, TALENs were shown to generate targeted DSBs first in yeast [15] and shortly thereafter in hiPSCs to stimulate gene targeting [16]. Both ZFNs and TALENs have continued to evolve as gene editing tools and are still considered to have high specificity due to their strict requirement to bind in pairs before cleavage of the target site [17].

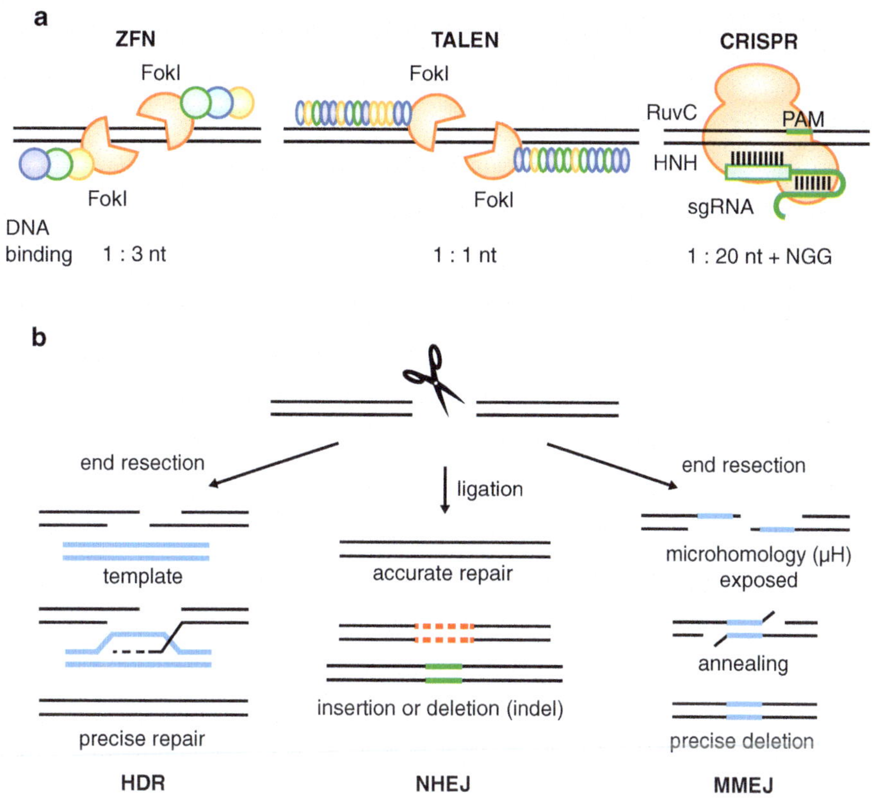

Fig. 7.1 Creation and repair of DNA double-strand breaks. (**a**) Schematic diagrams of ZFN, TALEN, and CRISPR programmable nucleases, indicating their unique DNA recognition systems and cleavage activities. (**b**) Three major repair outcomes mediated by cellular DSB repair machinery. HDR is a high-fidelity template-driven process, while NHEJ results in the formation of indels. MMEJ uses tandem homology, resulting in predictable deletions

The success of ZFNs and the more predictable binding properties of TALENs primed the field for a revolution in programmable DNA binding which arrived in 2012 in the form of CRISPR/Cas9 [18]. Particularly, the key effector molecule of the *Streptococcus pyogenes* (Sp) CRISPR antiviral adaptive immune system, SpCas9, has been most broadly adopted. In 2012 Jinek et al. characterized in vitro SpCas9 cleavage activity and engineered its associated crRNA and tracrRNA molecules—required for DNA binding specificity of the ribonucleoprotein (RNP) complex (Fig. 7.1a)—to generate a single-guide RNA or sgRNA [18]. Through a variable 20 nt region of the sgRNA sequence, the SpCas9-sgRNA RNP can be made to recognize a corresponding target genomic region where the only requirement is a downstream "NGG" sequence known as the protospacer adjacent motif (PAM). Recognition of the 20 nt target site and PAM was shown to induce DNA cleavage in mammalian cells, as well as stimulate gene targeting [19, 20]. These initial developments helped to democratize programmable nucleases through simplicity of design and application. Since then, SpCas9 has maintained a strong lead among a plethora of newly discovered bacterial endonucleolytic ribonucleoproteins, each with their own unique variations on target site recognition, gRNA structure, PAM requirements, nuclease domains, and DNA or RNA cleavage patterns [21]. Beyond this, SpCas9 continues to be engineered extensively for properties such as improved on-target and reduced off-target activity [22], as well as broader PAM specificity [23–25], solidifying its lead role in genome editing.

7.3 Repairing Genomic DSBs

Engineered programmable nucleases are intended to act through stimulation of endogenous DSB repair (DSBR) pathways in order to enhance genome editing. The molecular mechanisms of DSB recognition and repair are context-dependent, varying among cell types and states. The known molecular mechanisms of DSBR are described at length in [26]. Relevant to the goals of this chapter, the endogenous responses to a genomic DSB may be categorized by their outcomes, suggesting three major pathways to DSB resolution (Fig. 7.1b).

Homology-directed repair (HDR) is a process that employs DNA templates homologous to sequences flanking the DSB to faithfully repair the broken chromosome (Fig. 7.1b, left) [26]. This is the core principle of homologous recombination (HR) for classic gene targeting [27–29] and typically the desired pathway to achieve precision gene editing. Resolution of DSBs by HDR is proceeded by recognition and end resection mediated by the Mre11-Rad50-Nbs1 (MRN) endo-/exonuclease complex, initiating a process in which further end processing by Exo1 and CtIP is essential [30]. RPA binds exposed regions of single-stranded DNA (ssDNA), and Rad51 promotes strand invasion for incorporation of the template sequence. Natively, the template for repair is the newly replicated sister chromatid, while in an experimental setting, exogenous template DNA is provided simultaneously with active nuclease.

Following the principles of classic gene targeting [27–29], custom template DNA for HDR is often provided as double-stranded DNA (dsDNA) in the form of a covalently closed circular plasmid (Fig. 7.2a). Alternatives include linear dsDNA fragments generated by restriction or PCR [31] or nicked circular templates [32] which have been demonstrated to stimulate higher HDR repair rates. Mimicking processed HDR intermediates, long ssDNA generated by reverse transcription of RNA, or strand-specific DNA digestion has also been shown to improve targeting [33]. Obviating the need for cloning or enzymatic preparation, and therefore popular for their simplicity, synthetic single-stranded oligodeoxynucleotides (ssODNs) of ~100–200 bp in length have been employed extensively as HDR templates in multiple cell types, including hiPSCs [34–38]. While most studies opt to design ssODNs with symmetric homology across the DSB site, some suggest that asymmetry and strand specificity may be more compatible with repair of SpCas9-generated DSBs by improving annealing and invasion [39]. A second study

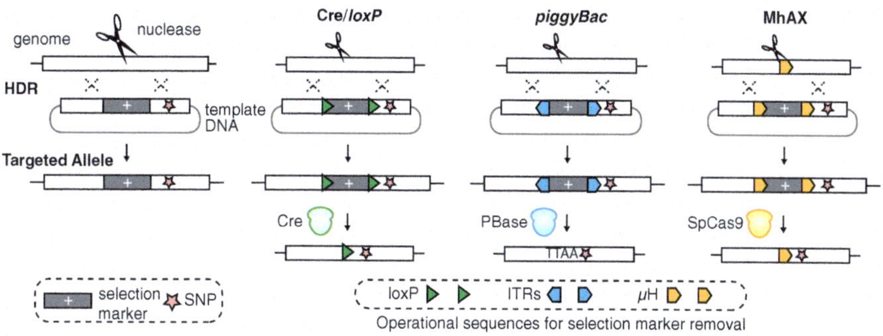

Fig. 7.2 Strategies for the removal of selection markers from the genome. Selection markers encoding antibiotic resistance genes or fluorescent proteins are effective to enrich for HDR events yet require additional strategies for their subsequent removal. Two standard methods, Cre/*loxP* and *piggyBac* transposition, leave behind traces of exogenous sequence in the genome. A novel method, MhAX, employs duplication of endogenous sequences for traceless removal of transgenes

examining the phenomenon of asymmetry found that short, 30 bp 3′ arms were most efficient, irrespective of the strand [40]. In fact, longer dsDNA templates generated with similar 30 bp ssDNA overhangs improved rates of large modifications in both HEK293 and hiPSCs, implying a common repair pathway for the two template types.

Nonhomologous end-joining (NHEJ) is a repair outcome attributed to the recognition of DSBs by Ku 70/80, a protein complex expected to prevent MRN-mediated end resection and rather promote religation of broken ends by DNA ligase IV. Unlike HDR, NHEJ has no requirement for homology and no mechanism to reconstitute the loss of sequence information from DSB ends (Fig. 7.1b, center). Although accurate repair is possible, NHEJ is generally accepted to be an error-prone pathway, resulting in random insertions and deletions (indels) of various sizes through the joining of imperfect DSBs following nuclease cleavage. Thus, NHEJ can be a desirable outcome for gene disruption by creating knockout alleles using single or paired guide RNAs [41]. However, due to the formation of indels, NHEJ is often considered to generate nonproductive outcomes competing with the precision of HDR.

Microhomology-mediated end joining (MMEJ) represents a third prominent repair outcome (Fig. 7.1b, right). MMEJ is thought to occur when, following MRN- and CtIP-mediated end resection, short tandem homologous sequences of a few or more bases in length are exposed within the ssDNA regions on either side of the DSB. If the resulting microhomologous ssDNA is not bound by RPA [42], they pair to direct DSBR in a DNA synthesis-dependent manner involving polymerase theta (POLQ) [43]. The resulting mutation is a precise deletion of one copy of the microhomology along with the intervening sequence (Fig. 7.1b, right). Microhomology has been associated with the formation of human pathogenic deletions and translocations [44]. Reproducible deletions have been observed following CRISPR/Cas9 treatment of multiple cell types in culture [45], implicating the surrounding genomic DNA context at particular target loci in determining repair outcomes. Systematic analyses using artificial or endogenous loci with microhomologies have verified a prominent role of MMEJ in defining DSBR outcomes [46] and established that MMEJ is an especially prevalent pathway in hESCs and hiPSCs [47]. Thus, the sequence context of a DSB has a significant impact on the repair outcome, knowledge of which may be used to select appropriate nuclease target sites in order to bias DSBR outcomes toward particular types of mutations [48, 49].

7.4 Modulation of DSBR

Beyond the local sequence context, the cell state and timing of nuclease expression can influence the choice of DSBR outcomes. For example, the activity of MRN and therefore HDR is known to be closely linked to the cell cycle [26], and a high proportion of hESCs and hiPSCs are found in S phase [50], potentially explaining the normally high levels of HDR and MMEJ seen in pluripotent cells. Further enhancing this phenomenon, synchronization with nocodazole at M phase prior to transfection was seen to marginally improve repair rates in HEK293T cells and hESCs [51]. Others have made use of chemical inhibition of NHEJ components [52] or protein overexpression [53, 54] with the intention to bias repair toward HDR. While providing measurable improvements to HDR outcomes, these approaches should be used with caution, as global impairment of DSBR may compromise overall genomic stability.

Protein engineering of SpCas9 that limits or augments its activity provides a more refined approach to regulate DSBR outcomes at the site of nuclease cleavage. For example, posttranslational regulation of SpCas9 activity during the cell cycle was achieved by fusing SpCas9 to the protein Geminin [55]. Geminin is selectively degraded in G1 phase, such that SpCas9 activity peaked during S-G2-M phases, supporting nearly twofold higher levels of HDR. Two strategies aimed to accumulate repair machinery at the site of nuclease cleavage through either direct fusion of a CtIP sub-fragment to SpCas9 [56] or recruitment of an MS2-CtIP fusion protein to MS2 RNA loops in modified sgRNA [57]. In either case, the levels of on-target HDR or MMEJ were reported to be higher than cleavage with SpCas9 alone. Variations on these approaches may make it possible to influence DSBR locally, without having global effects on genome stability.

7.5 Delivery of Nucleases to hiPSCs

As demonstrated in the first applications of CRISPR/Cas9 in mammalian cell culture, plasmid vectors remain one of the most common methods for delivery of polII-driven SpCas9 and polIII-driven sgRNA expression cassettes [19]. Moreover, plasmid vectors can be engineered to encode additional functions such as fluorescence or drug selection that are valuable for cell enrichment, as described in the section below. On the other hand, SpCas9/sgRNA RNP complex can be assembled in vitro from recombinant Cas9 protein and synthetic sgRNAs (or crRNA with tracrRNA), and high-quality components are available from multiple commercial sources. RNP delivery with or without a template DNA can be achieved using chemical transfection or electroporation, and optimized delivery protocols have been developed for various cell types including hESCs and hiPSCs [58]. Unlike plasmids, which continue to produce nuclease over a period of days, RNP transfection provides a finite amount of nuclease resulting in a more immediate and acute effect. For NHEJ-mediated knockout, RNP delivery has proven lucrative, and with optimization template-mediated HDR using RNP has shown promising results [40].

Transgenic hiPSC lines with constitutive or drug-regulable expression of SpCas9 are a potent alternative to transient transfection. For example, TALEN-mediated knock-in of a tetO promoter-driven SpCas9 along with the dox-responsive rtTA to the AAVS1 locus in hESC lines gave rise to an inducible gene editing system referred to as iCRISPR [59]. The iCRISPR system achieved gene knockouts up to 60% and gene editing with ssODN templates up to 10% [59] and was demonstrated to aid in the construction of fluorescent reporter alleles without drug selection in hESCs [60]. AAVS1 is considered a safe-harbor locus with reliable expression, and no phenotypic consequence is known for hetero- or homozygous transgene insertion [61]. Still, as integration into AAVS1 requires additional screening efforts to determine correctly targeted cell lines, others have taken a more tractable approach to transgenesis using *piggyBac* transposition for tetO::SpCas9 delivery [62]. Efficiencies of 40–50% for NHEJ-mediated gene knockout and 10–20% for HDR-mediated gene targeting with ssODNs were reported using *piggyBac* transgenesis. Moreover, removal of SpCas9 after gene editing was achieved by transient expression of the *piggyBac* transposase. Insertion of SpCas9 with *piggyBac* provides a more tractable system across most hiPSC lines. However, transposon copy numbers and integration sites are difficult to control [63], putting cells at risk of insertion mutagenesis and making subsequent removal a complex task. Thus, careful titration of *piggyBac* transposon DNA is recommended during the preparation of SpCas9 transgenic hiPSCs.

7.6 Enriching for Desirable Gene Editing Events

While programmable nucleases provided a significant improvement to human gene targeting, hiPSCs edited by HDR remain the minority within a population. This is particularly problematic for the creation of recessive alleles, which require modification of both maternal and paternal gene copies. Thus, significant effort is still required to identify hiPSCs edited by HDR, while excluding those with NHEJ- or MMEJ-mediated indel events at one or both alleles. Toward these ends, protocols for "direct" and "indirect" selection of HDR outcomes have been developed. "Direct" selection entails selecting for the editing event itself, such as sensitive phenotypic or molecular assays for detection of particular genotypes or knock-in of an antibiotic or fluorescent positive selection marker along with the desired mutation at the target site. On the other hand, "indirect" selection refers to selecting for hiPSCs with high levels of nuclease activity or selecting those with HDR events at measurable, secondary loci. Both direct and indirect methods of selection are particularly valuable for gene editing with ssODN templates, since the ssODN itself encodes no inherent capacity for selection, and the resulting HDR modification rarely generates a genomic change amenable to phenotypic selection. A variety of methods are outlined and discussed with examples below.

Genome-editing strategies for which delivery of the nuclease is achieved by transient expression from plasmids are often designed such that they also include expression cassettes for antibiotic of fluorescent selectable markers [19]. Here, selection for gene editing is indirect, selecting for transfected cells that express the selection marker and, by extension, the nuclease. In turn, untransfected cells that have a lower chance of being edited are removed from the population. Puromycin resistance is an effective choice for transient selection given the rapid death of non-resistant cells [64]. It is recommended, however, that each new hiPSC line be individually tested for their sensitivity to drug dosages ranging from 0 to 0.5 μg/mL over treatment periods of 1–3 days. Using this approach to select for ssODN-edited alleles at autosomal- and X-linked loci offered concomitant enrichment of both HDR and NHEJ. Yet, in nearly 30% of the cases, hiPSCs with HDR editing of one allele suffered NHEJ indels on the other, affirming that indirect selection is independent of DSBR outcome. Similarly, enrichment for non-specific nuclease activity can be made using transient expression of GFP fluorescence. Fluorescence enables FACS sub-fractionation of marker-positive cell populations based on intensity [65], which shows a high correlation between GFP brightness and NHEJ outcomes at target loci but no specific HDR enrichment.

It has been suggested that gene editing outcomes at secondary selectable loci may be used to indirectly select for similar outcomes at non-selectable target loci, a phenomenon dubbed "co-conversion." This was first demonstrated by generating mutations in *C. elegans* genes which confer behavior or morphology phenotypes, such as *rol-6*, *dpy-10*, or *unc-58*, coincidentally with HDR co-conversion of a second, non-selectable allele using ssODN, PCR products, or plasmids [66]. Editing of the target locus occurred in 14–84% of injected worms, although few HDR events were homozygous. Examples in mammalian cells include co-conversion of the HPRT locus by NHEJ-mediated knockout and resistance to 6-thioguanine [67], targeting of the AAVS1 locus by HDR-mediated gene trap knock-in resulting in antibiotic resistance [68], or HDR-mediated ssODN editing of ATP1A1 to confer resistance to ouabain by mimicking a naturally occurring resistance mutation [69]. All three methods enriched for mutations at the non-selectable target loci, although similar DSBR outcomes at both loci were not always observed. An important caveat of co-conversion is that it requires permanent modification of a second locus that, for *C. elegans*, can be segregated in the F2 generation by breeding. Co-converted hiPSCs, on the other hand, will always bear the additional genetic modification.

"Direct" selection can make use of molecular genotyping assays designed to detect rare gene editing events within a greater population of hiPSCs. Droplet digital PCR (ddPCR)-based detection systems have been designed to resolve the extent of NHEJ events [70] or in hiPSCs to pinpoint single nucleotide polymorphisms (SNPs) edited via ssODN templates [71]. In this latter method, allele-specific fluorescence-conjugated TaqMan probes that distinguish between the normal and mutant DNA sequences were employed to screen sub-pooled populations of hiPSCs. Positive sub-pools are then expanded, sub-fractionated, and rescreened in a process referred to as sib selection. While custom TaqMan probe design can present a technical challenge, ddPCR followed by single-cell cloning enables detection and isolation of edited hiPSCs without antibiotic selection, representing a sensitive screen for specific NHEJ knockout mutations or ssODN editing.

Returning to more classic gene targeting approaches [27–29], plasmid DNA templates with antibiotic or fluorescent markers allow direct positive/negative selection for HDR and against random integration events, respectively [72, 73]. Plasmid templates require cloning of the homology arms and selection markers, yet the larger homology regions provide advantages over ssODNs including maintenance and targeting of specific allelic haplotypes [74], larger conversion tracts [75], and dual selection markers for the isolation of homozygous targeted cells [76]. Yet, hiPSCs enriched by selection markers will retain the transgene at the target locus, which may result in pleiotropic effects on neighboring genes [77]. Therefore, removal of selection markers is a necessary step following the identification of targeted clones, and various approaches have been developed to address this issue.

7.7 Precise Removal of Selection Markers

Strategies for selection marker removal following gene targeting are illustrated in Fig. 7.2. By far the most common approach employs Cre/*loxP*, a site-specific recombination system effective across various species and cell types including hESCs [78] and hiPSCs [79]. Demonstrations of Cre-mediated selection marker removal in gene editing to derive disease models include Parkinson's disease [38], cystic fibrosis [80], sickle cell anemia [81], epidermolysis bullosa [82], and Alzheimer's disease [83], among many others. Although Cre excision is efficient and high-fidelity, a single 34 bp *loxP* site is always retained at the target site. Since residual *loxP* sites have been shown to influence target gene expression [84], positioning of *loxP*-flanked selection markers must be done with careful consideration.

Unique among mobile DNA elements, the *piggyBac* transposon is exquisitely precise in its excision from "TTAA" tetranucleotide sequences in donor DNA [85] or the genome of mouse and human iPSCs [63]. Applied to gene targeting, selection markers are flanked with 5′ and 3′ *piggyBac* inverted terminal repeats (ITRs) and cloned into plasmid template homology arms nearby designer mutations, in a manner comparable to Cre/*loxP* (Fig. 7.2). Following targeting, *piggyBac* selection marker excision is achieved by transient expression of native *piggyBac* transposase (PBase) or hyperactive PBase (hyPBase) [86]. This approach has been broadly adopted for hiPSC gene editing, for example, in the correction of A1AT [87], SOD1

[88], or HBB [89] mutations. Importantly, variable efficiencies of *piggyBac* excision from different genomic loci have been observed [63], presumably the cause of undefined genetic or epigenetic factors. Moreover, reintegration of the *piggyBac* transposon can occur following excision, which may be partially mitigated by engineered excision-competent/integration-defective ($Exc^{+}Int^{-}$) PBase [90]. Still, *piggyBac*-mediated genome editing is limited by a requirement for endogenous or engineered TTAA sites within close proximity to designer mutations.

The above examples bring to question the possibility of removing exogenous DNA from the genome without residual operational sequences, in a completely scarless manner. Conceptually similar to recombinase or transposase operational sequences, the reproducible formation of precise deletions by MMEJ (Fig. 7.1b, right) is also dependent upon short DNA duplications flanking the region to be removed [26, 44]. During the evaluation of various TALEN protein architectures in hiPSCs, we made the serendipitous observation that DSB formation within HPRT exon3 reproducibly resulted in a deletion of 17 bp, where the position 4 bp upstream of the deletion site was variable, either an A or T base [91, 92]. Interestingly, these same deletion mutations were observed previously in HEK293T cells [93], although their prevalence was not a subject of discussion. Further analysis of the common 17 bp deletion in iPSCs revealed that the TALEN target site is flanked by an imperfect 9 bp microhomology we called µ5W3, where W is the degenerate A/T base [91].

Through this observation we devised a new method called *m*icro*h*omology-*a*ssisted e*x*cision or MhAX [91]. MhAX is based on two core principles: (1) that MMEJ directs precise deletion of one copy of the microhomology along with the intervening sequence and (2) that an imperfect microhomology results in one of two possible allelic outcomes. Targeting the APRT gene in order to recreate the APRT*J disease allele [94], we engineered the circular plasmid templates to have duplicated genomic sequences (microhomologies) on either side of a selection marker (Fig. 7.2) [91]. To create duplicated microhomology from singlet endogenous sequences, the left and right homology arms were simply extended in the 3′ and 5′ orientations, respectively, resulting in an overlap of ~30 bp of sequence. The overlap encompassed the genomic target site, including the position of the APRT*J disease mutation. Additionally, protospacer sequences for SpCas9 without cognate targets in the human genome were nested between the engineered microhomology and selection marker in a divergent orientation (PAM-out). This orientation meant that during SpCas9-mediated cassette excision, DSB formation would occur at the most distal ends, 3 bp upstream of the outwardly oriented PAMs and proximal to the engineered microhomology, stimulating MhAX (Fig. 7.2). Direct selection at the targeting and excision steps employed both antibiotic and fluorescent markers to enable FACS enrichment. MhAX occurred in 35% of heterozygous targeted hiPSCs and at both alleles in homozygous hiPSCs at a rate of ~11%. The non-MMEJ alleles were ~2/3 random indels and 1/3 perfect NHEJ repair of the two flanking cut sites. This data proves that endogenous sequences can be employed through duplication as engineered microhomology, which can then be resolved with precision through endogenous DSBR pathways. Moreover, intentionally engineering imperfect microhomology led to two possible outcomes at each repaired allele: either normal or mutant. Thus, MhAX represents a novel approach to generate an allelic panel of hiPSCs from a single parental hiPSC clone [91].

7.8 Considering Individuality in hiPSCs Genome Editing

Celebrating over 10 years of technology development and application [1], reprogramming of somatic cells to hiPSCs has become a routine procedure in many laboratories. hiPSCs have been generated and banked from multitudes of patients [3] representing various genetic backgrounds such as HLA homozygosity [95] or polymorphic SNP alleles [96]. Of particular value for these resources is the potential to directly derive isogenic controls, where genetic background is invariable apart from the mutation of interest. Pairs of isogenic hiPSC have been derived from monozygotic twins, for example, in the characterization of neurodevelopmental defects related to trisomy 21 where the twins were discordant for trisomy [97]. As such cases are extraordinarily rare, the application of programmable nucleases to correct patient hiPSCs or reproduce putative disease mutations in hiPSCs with otherwise healthy genetic backgrounds is a common approach as it theoretically leads to perfect isogenic pairs of hiPSC clones.

Human genetic variation can compromise safe and predictable applications of programmable nucleases in hiPSCs but can also provide unique opportunities for precision editing. A fundamental complication arising from haplotype diversity is in the prediction of potential off-target sites which are similar in sequence to the intended target and may therefore be cleaved by programmable nucleases [98]. Analysis of thousands of whole genome sequences compiled from the 1000 Genomes Project, gnomAD database, and French Canadian dataset revealed that SNPs at disease-relevant loci can reduce on-target and exasperate off-target cleavage potential of CRISPR/Cas9 cleavage [99]. These calculations suggest that for more sensitive applications, techniques such as whole genome sequencing or in vitro cleavage analysis [100] using target hiPSC genomic DNA may be required to predict off-target loci for individuals. Although genetic variation in the human population can be confounding, it can also be used to drive unique genetic outcomes. Examples of allele-specific gene disruption by NHEJ using sgRNAs designed against heterozygous PAM-generating SNPs [101, 102] hold promise for unique therapeutic strategies and improved specificity in targeting particular allele haplotypes. Clearly, the heterozygous genetic backgrounds of hiPSCs require special considerations compared to most standard genetic models.

7.9 Conclusions

Human iPSCs, given their personal nature and unique in vitro properties of clonal growth and capacity for differentiation, provide an important access point to human functional genomics and future personal therapies. Combined with modern gene editing technologies, particular genotypes can be created and tested in various genetic backgrounds to assess disease predisposition or corrected in patient cells resulting in autologous cell resources for transplant. Future improvements in programmable endonuclease-mediated genome editing will undoubtedly address outstanding issues impacting the identification of and propensity for on- and off-target cleavage, as well as the predictably of editing outcomes. Accumulating knowledge of both CRIPSR/Cas9 biology and DSBR pathways in mammalian cells will drive improvements in these areas. Finally, gene editing without DSB formation, a process known as "base editing" that works via in situ base modification by enzymes such as cytidine deaminases [103–105], may eventually take a leading role in generating single-base changes with absolute precision. Still, it remains unlikely that base editing will make the programmable nuclease methods described herein obsolete, as each serves their particular purposes. Whatever shape future gene editing tools may have, they will continue to drive a revolution in human health, increasing the potential for medical applications of stem cells.

Acknowledgments I would like to acknowledge all those who have contributed to the rapid growth in the fields of induced pluripotency and gene editing and recognize their important works not cited here for the sake of brevity. This manuscript was funded in part by a grant to K.W. from the Japan Agency for Medical Research and Development (AMED, no. 18bm0804001h0002).

Competing Financial Interests The authors declare no competing financial interests.

References

1. Takahashi K, Yamanaka S. A decade of transcription factor-mediated reprogramming to pluripotency. Nat Rev Mol Cell Biol. 2016;17(3):183–93.
2. Clevers H. Modeling development and disease with organoids. Cell. 2016;165:1586–97.
3. Kim J-H, Kurtz A, Yuan B-Z, Zeng F, Lomax G, Loring JF, Crook J, Ju JH, Clarke L, Inamdar MS, et al. Report of the international stem cell banking initiative workshop activity: current hurdles and progress in seed-stock banking of human pluripotent stem cells. Stem Cells Transl Med. 2017;6:1956–62.
4. Sakuma T, Woltjen K. Nuclease-mediated genome editing: at the front-line of functional genomics technology. Develop Growth Differ. 2014;56:2–13.
5. Zeltner N, Studer L. Pluripotent stem cell-based disease modeling: current hurdles and future promise. Curr Opin Cell Biol. 2015;37:102–10.
6. Kim HS, Bernitz JM, Lee D-F, Lemischka IR. Genomic editing tools to model human diseases with isogenic pluripotent stem cells. Stem Cells Dev. 2014;23:2673–86.
7. Jasin M. Genetic manipulation of genomes with rare-cutting endonucleases. Trends Genet. 1996;12:224–8.

8. Pavletich NP, Pabo CO. Zinc finger-DNA recognition: crystal structure of a Zif268-DNA complex at 2.1 A. Science. 1991;252:809–17.
9. Kim YG, Cha J, Chandrasegaran S. Hybrid restriction enzymes: zinc finger fusions to Fok I cleavage domain. Proc Natl Acad Sci U S A. 1996;93:1156–60.
10. Smith J, Bibikova M, Whitby FG, Reddy AR, Chandrasegaran S, Carroll D. Requirements for double-strand cleavage by chimeric restriction enzymes with zinc finger DNA-recognition domains. Nucleic Acids Res. 2000;28:3361–9.
11. Elrod-Erickson M, Rould MA, Nekludova L, Pabo CO. Zif268 protein-DNA complex refined at 1.6 A: a model system for understanding zinc finger-DNA interactions. Structure. 1996;4:1171–80.
12. Hockemeyer D, Soldner F, Beard C, Gao Q, Mitalipova M, Dekelver RC, Katibah GE, Amora R, Boydston EA, Zeitler B, et al. Efficient targeting of expressed and silent genes in human ESCs and iPSCs using zinc-finger nucleases. Nat Biotechnol. 2009;27:851–7.
13. Boch J, Scholze H, Schornack S, Landgraf A, Hahn S, Kay S, Lahaye T, Nickstadt A, Bonas U. Breaking the code of DNA binding specificity of TAL-type III effectors. Science. 2009;326:1509–12.
14. Moscou MJ, Bogdanove AJ. A simple cipher governs DNA recognition by TAL effectors. Science. 2009;326:1501.
15. Christian M, Cermak T, Doyle EL, Schmidt C, Zhang F, Hummel A, Bogdanove AJ, Voytas DF. Targeting DNA double-strand breaks with TAL effector nucleases. Genetics. 2010;186:757–61.
16. Hockemeyer D, Wang H, Kiani S, Lai CS, Gao Q, Cassady JP, Cost GJ, Zhang L, Santiago Y, Miller JC, et al. Genetic engineering of human pluripotent cells using TALE nucleases. Nat Biotechnol. 2011;29(8):731–4.
17. Pattanayak V, Guilinger JP, Liu DR. Determining the specificities of TALENs, Cas9, and other genome-editing enzymes. Methods Enzymol. 2014;546:47–78.
18. Jinek M, Chylinski K, Fonfara I, Hauer M, Doudna JA, Charpentier E. A programmable dual-RNA-guided DNA endonuclease in adaptive bacterial immunity. Science. 2012;337:816–21.
19. Ran FA, Hsu PD, Wright J, Agarwala V, Scott DA, Zhang F. Genome engineering using the CRISPR-Cas9 system. Nat Protoc. 2013;8:2281–308.
20. Mali P, Yang L, Esvelt KM, Aach J, Guell M, DiCarlo JE, Norville JE, Church GM. RNA-guided human genome engineering via Cas9. Science. 2013;339(6121):823–6.
21. Koonin EV, Makarova KS, Zhang F. Diversity, classification and evolution of CRISPR-Cas systems. Curr Opin Microbiol. 2017;37:67–78.
22. Lee JK, Jeong E, Lee J, Jung M, Shin E, Kim Y-H, Lee K, Jung I, Kim D, Kim S, et al. Directed evolution of CRISPR-Cas9 to increase its specificity. Nat Commun. 2018;9:3048.
23. Hirano S, Nishimasu H, Ishitani R, Nureki O. Structural basis for the altered PAM specificities of engineered CRISPR-Cas9. Mol Cell. 2016;61:886–94.
24. Kleinstiver BP, Pattanayak V, Prew MS, Tsai SQ, Nguyen NT, Zheng Z, Joung JK. High-fidelity CRISPR-Cas9 nucleases with no detectable genome-wide off-target effects. Nature. 2016;529:490–5.
25. Kleinstiver BP, Prew MS, Tsai SQ, Topkar VV, Nguyen NT, Zheng Z, Gonzales APW, Li Z, Peterson RT, Yeh J-RJ, et al. Engineered CRISPR-Cas9 nucleases with altered PAM specificities. Nature. 2015;523:481–5.
26. Ceccaldi R, Rondinelli B, D'Andrea AD. Repair pathway choices and consequences at the double-strand break. Trends Cell Biol. 2016;26:52–64.
27. Mansour SL, Thomas KR, Capecchi MR. Disruption of the proto-oncogene int-2 in mouse embryo-derived stem cells: a general strategy for targeting mutations to non-selectable genes. Nature. 1988;336:348–52.
28. Smithies O, Gregg RG, Boggs SS, Koralewski MA, Kucherlapati RS. Insertion of DNA sequences into the human chromosomal beta-globin locus by homologous recombination. Nature. 1985;317:230–4.

29. Thomas KR, Folger KR, Capecchi MR. High frequency targeting of genes to specific sites in the mammalian genome. Cell. 1986;44:419–28.
30. Shibata A, Moiani D, Arvai AS, Perry J, Harding SM, Genois M-M, Maity R, van Rossum-Fikkert S, Kertokalio A, Romoli F, et al. DNA double-Strand break repair pathway choice is directed by distinct MRE11 nuclease activities. Mol Cell. 2014;53:7–18.
31. Song F, Stieger K. Optimizing the DNA donor template for homology-directed repair of double-strand breaks. Mol Ther Nucleic Acid. 2017;7:53–60.
32. Nakajima K, Zhou Y, Tomita A, Hirade Y, Gurumurthy CB, Nakada S. Precise and efficient nucleotide substitution near genomic nick via noncanonical homology-directed repair. Genome Res. 2018;28:223–30.
33. Miura H, Quadros RM, Gurumurthy CB, Ohtsuka M. Easi-CRISPR for creating knock-in and conditional knockout mouse models using long ssDNA donors. Nat Protoc. 2017;13:195–215.
34. Chen F, Pruett-Miller SM, Huang Y, Gjoka M, Duda K, Taunton J, Collingwood TN, Frodin M, Davis GD. High-frequency genome editing using ssDNA oligonucleotides with zinc-finger nucleases. Nat Meth. 2011;8:753–5.
35. Merkert S, Wunderlich S, Bednarski C, Beier J, Haase A, Dreyer A-K, Schwanke K, Meyer J, Göhring G, Cathomen T, et al. Efficient designer nuclease-based homologous recombination enables direct PCR screening for footprintless targeted human pluripotent stem cells. Stem Cell Reports. 2014;2:107–18.
36. Orlando SJ, Santiago Y, Dekelver RC, Freyvert Y, Boydston EA, Moehle EA, Choi VM, Gopalan SM, Lou JF, Li J, et al. Zinc-finger nuclease-driven targeted integration into mammalian genomes using donors with limited chromosomal homology. Nucleic Acids Res. 2010;38:e152.
37. Radecke S, Radecke F, Cathomen T, Schwarz K. Zinc-finger nuclease-induced gene repair with oligodeoxynucleotides: wanted and unwanted target locus modifications. Mol Ther. 2010;18:743–53.
38. Soldner F, Laganière J, Cheng AW, Hockemeyer D, Gao Q, Alagappan R, Khurana V, Golbe LI, Myers RH, Lindquist S, et al. Generation of isogenic pluripotent stem cells differing exclusively at two early onset parkinson point mutations. Cell. 2011;146:318–31.
39. Richardson CD, Ray GJ, DeWitt MA, Curie GL, Corn JE. Enhancing homology-directed genome editing by catalytically active and inactive CRISPR-Cas9 using asymmetric donor DNA. Nat Biotechnol. 2016;34(3):339–44.
40. Liang X, Potter J, Kumar S, Ravinder N, Chesnut JD. Enhanced CRISPR/Cas9-mediated precise genome editing by improved design and delivery of gRNA, Cas9 nuclease, and donor DNA. J Biotechnol. 2017;241:136–46.
41. Liu Z, Hui Y, Shi L, Chen Z, Xu X, Chi L, Fan B, Fang Y, Liu Y, Ma L, et al. Efficient CRISPR/Cas9-mediated versatile, predictable, and donor-free gene knockout in human pluripotent stem cells. Stem Cell Reports. 2016;7:496–507.
42. Deng SK, Gibb B, de Almeida MJ, Greene EC, Symington LS. RPA antagonizes microhomology-mediated repair of DNA double-strand breaks. Nat Struct Mol Biol. 2014;21:405–12.
43. Kent T, Chandramouly G, McDevitt SM, Ozdemir AY, Pomerantz RT. Mechanism of microhomology-mediated end-joining promoted by human DNA polymerase θ. Nat Struct Mol Biol. 2015;22:230–7.
44. McVey M, Lee SE. MMEJ repair of double-strand breaks (director's cut): deleted sequences and alternative endings. Trends Genet. 2008;24:529–38.
45. van Overbeek M, Capurso D, Carter MM, Thompson MS, Frias E, Russ C, Reece-Hoyes JS, Nye C, Gradia S, Vidal B, et al. DNA repair profiling reveals nonrandom outcomes at Cas9-mediated breaks. Mol Cell. 2016;63(4):633–46.
46. Allen FR, Crepaldi LR, Alsinet-Armengol C, Strong A, Kleshchevnikov V, De Angeli P, Palenikova P, Kosicki M, Bassett AR, Harding H, et al. Mutations generated by repair of Cas9-induced double strand breaks are predictable from surrounding sequence. bioRxiv. 2018:400341. https://doi.org/10.1101/400341.

47. Taheri-Ghahfarokhi A, Taylor BJM, Nitsch R, Lundin A, Cavallo A-L, Madeyski-Bengtson K, Karlsson F, Clausen M, Hicks R, Mayr LM, et al. Decoding non-random mutational signatures at Cas9 targeted sites. Nucleic Acids Res. 2018;46:8417–34.
48. Ata H, Ekstrom TL, Martínez-Gálvez G, Mann CM, Dvornikov AV, Schaefbauer KJ, Ma AC, Dobbs D, Clark KJ, Ekker SC. Robust activation of microhomology-mediated end joining for precision gene editing applications. PLoS Genet. 2018;14:e1007652.
49. Bae S, Kweon J, Kim HS, Kim J-S. Microhomology-based choice of Cas9 nuclease target sites. Nat Methods. 2014;11:705–6.
50. Hindley C, Philpott A. The cell cycle and pluripotency. Biochem J. 2013;451:135–43.
51. Lin S, Staahl BT, Alla RK, Doudna JA. Enhanced homology-directed human genome engineering by controlled timing of CRISPR/Cas9 delivery. elife. 2014;3:e04766.
52. Yu C, Liu Y, Ma T, Liu K, Xu S, Zhang Y, Liu H, La Russa M, Xie M, Ding S, et al. Small molecules enhance CRISPR genome editing in pluripotent stem cells. Cell Stem Cell. 2015;16:142–7.
53. Canny MD, Moatti N, Wan LCK, Fradet-Turcotte A, Krasner D, Mateos-Gomez PA, Zimmermann M, Orthwein A, Juang Y-C, Zhang W, et al. Inhibition of 53BP1 favors homology-dependent DNA repair and increases CRISPR–Cas9 genome-editing efficiency. Nat Biotechnol. 2017;36:95–102.
54. Chu VT, Weber T, Wefers B, Wurst W, Sander S, Rajewsky K, Kühn R. Increasing the efficiency of homology-directed repair for CRISPR-Cas9-induced precise gene editing in mammalian cells. Nat Biotechnol. 2015;33:543–8.
55. Gutschner T, Haemmerle M, Genovese G, Draetta GF, Chin L. Post-translational regulation of Cas9 during G1 enhances homology-directed repair. Cell Rep. 2016;14:1555–66.
56. Charpentier M, Khedher AHY, Menoret S, Brion A, Lamribet K, Dardillac E, Boix C, Perrouault L, Tesson L, Geny S, et al. CtIP fusion to Cas9 enhances transgene integration by homology-dependent repair. Nat Commun. 2018;9(1):1133.
57. Nakade S, Mochida K, Kunii A, Nakamae K, Aida T, Tanaka K, Sakamoto N, Sakuma T, Yamamoto T. Biased genome editing using the local accumulation of DSB repair molecules system. Nat Commun. 2018;9(1):3270.
58. Liang X, Potter J, Kumar S, Zou Y, Quintanilla R, Sridharan M, Carte J, Chen W, Roark N, Ranganathan S, et al. Rapid and highly efficient mammalian cell engineering via Cas9 protein transfection. J Biotechnol. 2015;208:44–53.
59. Gonzalez F, Zhu Z, Shi Z-D, Lelli K, Verma N, Li QV, Huangfu D. An iCRISPR platform for rapid, multiplexable, and inducible genome editing in human pluripotent stem cells. Cell Stem Cell. 2014;15(2):215–26.
60. Zhu Z, Verma N, Gonzalez F, Shi Z-D, Huangfu D. A CRISPR/Cas-mediated selection-free knockin strategy in human embryonic stem cells. Stem Cell Reports. 2015;4:1103–11.
61. Oceguera-Yanez F, Kim S-I, Matsumoto T, Tan GW, Xiang L, Hatani T, Kondo T, Ikeya M, Yoshida Y, Inoue H, et al. Engineering the AAVS1 locus for consistent and scalable transgene expression in human iPSCs and their differentiated derivatives. Methods. 2016;101:43–55.
62. Wang G, Yang L, Grishin D, Rios X, Ye LY, Hu Y, Li K, Zhang D, Church GM, Pu WT. Efficient, footprint-free human iPSC genome editing by consolidation of Cas9/CRISPR and piggyBac technologies. Nat Protoc. 2017;12:88–103.
63. Woltjen K, Michael IP, Mohseni P, Desai R, Mileikovsky M, Hämäläinen R, Cowling R, Wang W, Liu P, Gertsenstein M, et al. piggyBac transposition reprograms fibroblasts to induced pluripotent stem cells. Nature. 2009;458:766–70.
64. Steyer B, Bu Q, Cory E, Jiang K, Duong S, Sinha D, Steltzer S, Gamm D, Chang Q, Saha K. Scarless genome editing of human pluripotent stem cells via transient puromycin selection. Stem Cell Reports. 2018;10(2):642–54.
65. Lonowski LA, Narimatsu Y, Riaz A, Delay CE, Yang Z, Niola F, Duda K, Ober EA, Clausen H, Wandall HH, et al. Genome editing using FACS enrichment of nuclease-expressing cells and indel detection by amplicon analysis. Nat Protoc. 2017;12:581–603.

66. Arribere JA, Bell RT, Fu BXH, Artiles KL, Hartman PS, Fire AZ. Efficient marker-free recovery of custom genetic modifications with CRISPR/Cas9 in *Caenorhabditis elegans*. Genetics. 2014;198:837–46.
67. Liao S, Tammaro M, Yan H. Enriching CRISPR-Cas9 targeted cells by co-targeting the HPRT gene. Nucleic Acids Res. 2015;43:e134.
68. Mitzelfelt KA, McDermott-Roe C, Grzybowski MN, Marquez M, Kuo C-T, Riedel M, Lai S, Choi MJ, Kolander KD, Helbling D, et al. Efficient precision genome editing in iPSCs via genetic co-targeting with selection. Stem Cell Reports. 2017;8:491–9.
69. Agudelo D, Duringer A, Bozoyan L, Huard CC, Carter S, Loehr J, Synodinou D, Drouin M, Salsman J, Dellaire G, et al. Marker-free coselection for CRISPR-driven genome editing in human cells. Nat Methods. 2017;14(6):615–20.
70. Findlay SD, Vincent KM, Berman JR, Postovit L-M. A digital PCR-based method for efficient and highly specific screening of genome edited cells. PLoS One. 2016;11:e0153901.
71. Miyaoka Y, Chan AH, Judge LM, Yoo J, Huang M, Nguyen TD, Lizarraga PP, So P-L, Conklin BR. Isolation of single-base genome-edited human iPS cells without antibiotic selection. Nat Methods. 2014;11:291–3.
72. Capecchi MR. Gene targeting in mice: functional analysis of the mammalian genome for the twenty-first century. Nat Rev Genet. 2005;6:507–12.
73. Hatada S, Arnold LW, Hatada T, Cowhig JE, Ciavatta D, Smithies O. Isolating gene-corrected stem cells without drug selection. Proc Natl Acad Sci U S A. 2005;102:16357–61.
74. Deyle DR, Li LB, Ren G, Russell DW. The effects of polymorphisms on human gene targeting. Nucleic Acids Res. 2014;42:3119–24.
75. Yang L, Güell M, Byrne S, Yang JL, De Los Angeles A, Mali P, Aach J, Kim-Kiselak C, Briggs AW, Rios X, et al. Optimization of scarless human stem cell genome editing. Nucleic Acids Res. 2013;41:9049–61.
76. Iiizumi S, Nomura Y, So S, Uegaki K, Aoki K, Shibahara K-i, Adachi N, Koyama H. Simple one-week method to construct gene-targeting vectors: application to production of human knockout cell lines. BioTechniques. 2006;41:311–6.
77. Pham CT, MacIvor DM, Hug BA, Heusel JW, Ley TJ. Long-range disruption of gene expression by a selectable marker cassette. Proc Natl Acad Sci U S A. 1996;93:13090–5.
78. Araki K, Imaizumi T, Okuyama K, Oike Y, Yamamura K. Efficiency of recombination by Cre transient expression in embryonic stem cells: comparison of various promoters. J Biochem. 1997;122:977–82.
79. Davis RP, Costa M, Grandela C, Holland AM, Hatzistavrou T, Micallef SJ, Li X, Goulburn AL, Azzola L, Elefanty AG, et al. A protocol for removal of antibiotic resistance cassettes from human embryonic stem cells genetically modified by homologous recombination or transgenesis. Nat Protoc. 2008;3:1550–8.
80. Crane AM, Kramer P, Bui JH, Chung WJ, Li XS, Gonzalez-Garay ML, Hawkins F, Liao W, Mora D, Choi S, et al. Targeted correction and restored function of the CFTR gene in cystic fibrosis induced pluripotent stem cells. Stem Cell Reports. 2015;4:569–77.
81. Zou J, Mali P, Huang X, Dowey SN, Cheng L. Site-specific gene correction of a point mutation in human iPS cells derived from an adult patient with sickle cell disease. Blood. 2011;118:4599–608.
82. Sebastiano V, Zhen HH, Haddad B, Derafshi BH, Bashkirova E, Melo SP, Wang P, Leung TL, Siprashvili Z, Tichy A, et al. Human COL7A1-corrected induced pluripotent stem cells for the treatment of recessive dystrophic epidermolysis bullosa. Sci Transl Med. 2014;6:264ra163.
83. Kondo T, Imamura K, Funayama M, Tsukita K, Miyake M, Ohta A, Woltjen K, Nakagawa M, Asada T, Arai T, et al. iPSC-based compound screening and in vitro trials identify a synergistic anti-amyloid β combination for Alzheimer's disease. Cell Rep. 2017;21(8):2304–12.
84. Meier ID, Bernreuther C, Tilling T, Neidhardt J, Wong YW, Schulze C, Streichert T, Schachner M. Short DNA sequences inserted for gene targeting can accidentally interfere with off-target gene expression. FASEB J. 2010;24:1714–24.

85. Li X, Lobo N, Bauser CA, Fraser MJ Jr. The minimum internal and external sequence requirements for transposition of the eukaryotic transformation vector piggyBac. Mol Gen Genomics. 2001;266:190–8.
86. Yusa K, Zhou L, Li MA, Bradley A, Craig NL. A hyperactive piggyBac transposase for mammalian applications. Proc Natl Acad Sci. 2011;108:1531–6.
87. Yusa K, Rashid ST, Strick-Marchand H, Varela I, Liu P-Q, Paschon DE, Miranda E, Ordóñez A, Hannan NRF, Rouhani FJ, et al. Targeted gene correction of α1-antitrypsin deficiency in induced pluripotent stem cells. Nature. 2011;478:391–4.
88. Imamura K, Izumi Y, Watanabe A, Tsukita K, Woltjen K, Yamamoto T, Hotta A, Kondo T, Kitaoka S, Ohta A, et al. The Src/c-Abl pathway is a potential therapeutic target in amyotrophic lateral sclerosis. Sci Transl Med. 2017;9(391):eaaf3962.
89. Xie F, Ye L, Chang JC, Beyer AI, Wang J, Muench MO, Kan YW. Seamless gene correction of β-thalassemia mutations in patient-specific iPSCs using CRISPR/Cas9 and piggyBac. Genome Res. 2014;24:1526–33.
90. Li X, Burnight ER, Cooney AL, Malani N, Brady T, Sander JD, Staber J, Wheelan SJ, Joung JK, McCray PB, et al. piggyBac transposase tools for genome engineering. Proc Natl Acad Sci. 2013;110:E2279–87.
91. Kim S-I, Matsumoto T, Kagawa H, Nakamura M, Hirohata R, Ueno A, Ohishi M, Sakuma T, Soga T, Yamamoto T, et al. Microhomology-assisted scarless genome editing in human iPSCs. Nat Commun. 2018;9:939.
92. Sakuma T, Hosoi S, Woltjen K, Suzuki K-I, Kashiwagi K, Wada H, Ochiai H, Miyamoto T, Kawai N, Sasakura Y, et al. Efficient TALEN construction and evaluation methods for human cell and animal applications. Genes Cells. 2013;18:315–26.
93. Cermak T, Doyle EL, Christian M, Wang L, Zhang Y, Schmidt C, Baller JA, Somia NV, Bogdanove AJ, Voytas DF. Efficient design and assembly of custom TALEN and other TAL effector-based constructs for DNA targeting. Nucleic Acids Res. 2011;39:e82.
94. Kamatani N, Kuroshima S, Hakoda M, Palella TD, Hidaka Y. Crossovers within a short DNA sequence indicate a long evolutionary history of the *APRT*J* mutation. Hum Genet. 1990;85(6):600–4.
95. Lee S, Huh JY, Turner DM, Lee S, Robinson J, Stein JE, Shim SH, Hong CP, Kang MS, Nakagawa M, et al. Repurposing the cord blood Bank for haplobanking of HLA-homozygous iPSCs and their usefulness to multiple populations. Stem Cells. 2018;126:663.
96. Warren CR, O'Sullivan JF, Friesen M, Becker CE, Zhang X, Liu P, Wakabayashi Y, Morningstar JE, Shi X, Choi J, et al. Induced pluripotent stem cell differentiation enables functional validation of GWAS variants in metabolic disease. Cell Stem Cell. 2017;20:547–57.
97. Hibaoui Y, Grad I, Letourneau A, Sailani MR, Dahoun S, Santoni FA, Gimelli S, Guipponi M, Pelte MF, Béna F, et al. Modelling and rescuing neurodevelopmental defect of Down syndrome using induced pluripotent stem cells from monozygotic twins discordant for trisomy 21. EMBO Mol Med. 2014;6:259–77.
98. Zhang X-H, Tee LY, Wang X-G, Huang Q-S, Yang S-H. Off-target effects in CRISPR/Cas9-mediated genome engineering. Mol Ther Nucleic Acid. 2015;4:e264.
99. Lessard S, Francioli L, Alfoldi J, Tardif J-C, Ellinor PT, MacArthur DG, Lettre G, Orkin SH, Canver MC. Human genetic variation alters CRISPR-Cas9 on- and off-targeting specificity at therapeutically implicated loci. Proc Natl Acad Sci. 2017;114:E11257–66.
100. Kim D, Bae S, Park J, Kim E, Kim S, Yu HR, Hwang J, Kim J-I, Kim J-S. Digenome-seq: genome-wide profiling of CRISPR-Cas9 off-target effects in human cells. Nat Methods. 2015;12:237–43.
101. Christie KA, Courtney DG, DeDionisio LA, Shern CC, Majumdar S, Mairs LC, Nesbit MA, Moore CBT. Towards personalised allele-specific CRISPR gene editing to treat autosomal dominant disorders. Sci Rep. 2017;7(1):16174.
102. Smith C, Abalde-Atristain L, He C, Brodsky BR, Braunstein EM, Chaudhari P, Jang Y-Y, Cheng L, Ye Z. Efficient and allele-specific genome editing of disease loci in human iPSCs. Mol Ther. 2015;23:570–7.

103. Hess GT, Tycko J, Yao D, Bassik MC. Methods and applications of CRISPR-mediated base editing in eukaryotic genomes. Mol Cell. 2017;68:26–43.
104. Komor AC, Kim YB, Packer MS, Zuris JA, Liu DR. Programmable editing of a target base in genomic DNA without double-stranded DNA cleavage. Nature. 2016;533:420–4.
105. Nishida K, Arazoe T, Yachie N, Banno S, Kakimoto M, Tabata M, Mochizuki M, Miyabe A, Araki M, Hara KY, et al. Targeted nucleotide editing using hybrid prokaryotic and vertebrate adaptive immune systems. Science. 2016;353(6305):aaf8729.

Part IV
iPS Applications

Chapter 8
Induced Pluripotent Stem Cell-Based Cell Therapy of the Retina

Seiji Takagi, Michiko Mandai, Yasuhiko Hirami, Yasuo Kurimoto, and Masayo Takahashi

Abstract This review outlines the progress and current knowledge of stem cell-based treatments for the retina. Basic research, which began as a study of neural stem cells in the 1980s, is now being applied for the clinical use of embryonic stem (ES) cells and induced pluripotent stem cells (iPSCs). The author's group had transplanted iPS cell-derived retinal pigment epithelial (RPE) cell sheets to the eye of a patient with exudative age-related macular degeneration (AMD) in 2014 as a clinical research. Since it was the first clinical study using iPS cell-derived cells, the safety and tumorigenicity of iPSCs products were especially a great concern. At 1 year after surgery, the transplanted sheet remained intact, and no serious adverse event including tumor formation had been observed.

Autologous iPSC, which have lower chance for tissue rejection, is an ideal cell source for a transplant therapy; however, production of autologous iPSC lines is too costly. To address this issue, we are currently exploring the use of allogenic RPE cells from iPSCs of HLA homozygote donors with the Center for iPSC Research and Application (CiRA) of Kyoto University.

Regarding photoreceptor implantation, we recently demonstrated a direct contact between the host bipolar cell-end and the presynaptic terminal of the graft pho-

S. Takagi
Department of Ophthalmology, Kobe City Medical Center General Hospital, Kobe, Japan

Department of Translational Research Division of Ophthalmology, Institute of Biomedical Research and Innovation, Kobe, Japan

Department of Ophthalmology, Toho University Ohashi Medical Center, Tokyo, Japan

M. Mandai · M. Takahashi (✉)
Department of Ophthalmology, Kobe City Medical Center General Hospital, Kobe, Japan

RIKEN Center for Developmental Biology, Kobe, Japan
e-mail: retinalab@cdb.riken.jp

Y. Hirami · Y. Kurimoto
Department of Ophthalmology, Kobe City Medical Center General Hospital, Kobe, Japan

Department of Translational Research Division of Ophthalmology, Institute of Biomedical Research and Innovation, Kobe, Japan

H. Inoue, Y. Nakamura (eds.), *Medical Applications of iPS Cells*, Current Human Cell Research and Applications,
https://doi.org/10.1007/978-981-13-3672-0_8

toreceptor by directly implanting retinal tissues derived from murine iPSC into a mouse model of end-stage retinal degeneration. We believe that our findings established a new proof of concept for transplanting ESC/iPSC retinas to restore vision in patients with end-stage retinal degeneration.

Keywords Autologous transplants · Clinical research · iPS cells · Retinal cells

8.1 Introduction

The neural retina is a component of the central nervous system (CNS); it comprises three layers of neural cells within the retina (photoreceptor cells, bipolar cells, and ganglion cells) and the retinal pigment epithelium (RPE) under the photoreceptors. Light perception from the photoreceptor—the innermost part of the retina—is transduced to the occipital cortex of the brain via bipolar cells, ganglion cells, and the optic nerve.

The RPE is a monolayer of cuboid cells that lies between Bruch's membrane and the neural retina and is essential for retinal homeostasis. Degeneration of the neural retina and/or RPE caused by certain diseases, such as age-related macular degeneration (AMD) and retinitis pigmentosa (RP), limits the efficacy of current therapies. Thus, regenerative stem cell therapy to replace the atrophic retinal components has received increasing attention in recent years. Although basic studies that explored its viability have been carried out since the 1980s, this treatment has only been applied in the clinical setting in recent years.

To date, there are two potential cell sources for transplantation therapies—embryonic stem cells (ESCs) [1] and induced pluripotent stem cells (iPSCs) [2–4]. We and others have reported the differentiation of retinal cells from both mouse and human ESCs and iPSCs [5–7], and these stem cell-derived retinal cells have opened a new avenue in retinal regenerative medicine.

In 2014, we performed a clinical study of transplanted autologous iPSC-derived retinal pigment epithelial cell sheets to the eye of a patient with exudative macular AMD [8]; this was the first clinical study of iPSC use in human. A study of iPSC-RPE allotransplantation in HLA-matched AMD patients is currently under way starting in 2017.

In this chapter, we describe the aforementioned clinical studies, as well as previous efforts leading to the implementation of the clinical studies. We also discuss future directions for clinical studies, including that of photoreceptor transplantation.

8.2 Neural Stem Cells for Retinal Regenerative Medicine

In the 1980s, the existence of stem cells, which were originally studied mainly in the context of blood cells, began to be verified in other biological systems. Using developing mouse retina, Cepko et al. demonstrated that neural stem cells can differentiate into all types of retinal cells including Müller glial cells [9], indicating that multipotent precursors are present in the developing CNS. In the 1990s, various studies reported the existence of neural progenitors in the sub-ventricular zone [10] and hippocampus [11].

Several studies reported that the generation of CNS cells can be influenced by growth factors in the culture of embryonic and adult mouse neural stem cells. Weiss et al. reported that the application of epidermal growth factor to spheres of undifferentiated cells induced the proliferation of two progenitor cell types, neurons and astrocytes [12]. Furthermore, Gage et al. reported that stem cell-like multipotent progenitors can be isolated from adult rat hippocampus using fibroblast growth factor [13].

In 1998, Takahashi and Gage demonstrated the plasticity of adult rat hippocampus-derived neural progenitor cells that were grafted into the vitreous cavity of adult and newborn rat eyes and their subsequent widespread integration and survival [14]. This study was the first stepping stone toward the development of stem cell-based treatments for the eye. Moreover, Akagi and Takahashi reported the culture of retinal progenitor cells isolated from embryonic rat retina in serum-free medium in the presence of both epidermal growth factor and basic fibroblast growth factor; these cells subsequently differentiated into neuron and glial cells, including rhodopsin-positive photoreceptor cells [15]. However, they appeared to lose their tissue specificity after an extended culture period.

The use of autologous neural stem cells for regenerative medicine has few ethical challenges and minimal chance of tissue rejection. However, the technique involves invasive tissue collection and difficulties in differentiating the cells to obtain sufficient number of cells; therefore, an alternative cell source was needed. Studies that assessed the differentiation of human pluripotent stem cells to RPE and retinal progenitor cells, initiated in the early 2000s, provided a theoretical base of retinal cell source for the treatment of retinal degeneration.

8.3 RPE Replacement Therapies for Age-Related Macular Degeneration

8.3.1 Role of the RPE

The RPE is a monolayer of pigmented cells that forms a blood-retinal barrier between the retinal photoreceptor and choroidal vessel layers. It has multiple functions, including regeneration of bleached opsins, phagocytosis, secretory activity, light absorption, transepithelial transport, a component of the visual cycle, and protection against photooxidation [16–18]. The RPE forms a unit with photoreceptor for these functions, and a loss of integrity of this unit promotes apoptosis or degeneration of the photoreceptor. Additionally, the RPE plays an important role in maintaining an immunosuppressive microenvironment in the subretinal space by secreting pigment epithelium-derived factor (PEDF) and transforming growth factor-beta (TGF-β) [19, 20]. This environment allows transplanted cells to survive.

8.3.2 Treatment Concept

Age-related macular degeneration (AMD) is a common disease that can reduce vision quality [21]. In patients with AMD, the center of the visual field is largely impaired, causing difficulties in recognizing faces or letters or performing daily activities. AMD is divided into the dry and wet types based on the presence of neovascularization growing upward from the choroid to the RPE and neural retina, with accompanying intra- and subretinal fluids and hemorrhage [22]. An intravitreal injection of anti-vascular endothelial growth factor (VEGF) is currently the first-line therapy for wet AMD, which is considerably more prevalent than the dry type in Japan [23]. The efficacy and safety for this therapy have been demonstrated in several large clinical trials [24, 25]. These treatments can prevent exudative changes from neovascularization, and frequent treatments can preserve visual acuity. However, this strategy does not target the underlying pathogenesis and cannot repair the sensory retina and RPE once degenerated.

Although AMD is a multifactorial disease influenced by genetic [26, 27], environmental, and lifestyle factors [28], pathological studies have shown that physical disruption and functional impairment of the RPE occur during the course of wet AMD [29]. Therefore, transplantation of sheets or suspension of healthy RPE cells for AMD patients has been attempted as a therapeutic option in the last two decades. Fetal or adult donor eye tissues have been considered for the source of RPE [30], but these are limited in quantity and may cause tissue rejection. Although improvement of visual function has been reported in some patients, the use of autologous peripheral RPE with or without the choroid involves significant surgical risk including massive hemorrhage and retinal detachment [31]. Moreover, as mentioned earlier, these cells are limited in quantity, which renders them less than ideal for the source of cells.

These facts suggested that a healthier and more abundant RPE source is required to render RPE transplantation a viable treatment option. Differentiating pluripotent stem cells to RPE is likely to provide a sufficient supply of viable RPE cells without genetic defects inherent to the AMD patient.

8.3.3 ESCs as a Source for RPE

ESCs, first cultured in 1998 by Thomson et al., are pluripotent stem cells isolated from the inner cell mass of a blastocyst [1]. Owing to their capacity for self-renewal, ESCs allow an unlimited supply of RPE and photoreceptor cells for replacement therapies. Numerous studies by our group and others have demonstrated the induction of retinal cells, including RPE cells, from ESCs in vitro.

Kawasaki et al. [32] were the first to report the differentiation of RPE-like cells from monkey ESCs. They reported large patches of pigmented Pax6-positive (an optic cup marker) hexagonal epithelium cells cultured from ESCs using the stromal cell-derived inducing activity (SDIA) method, which was originally developed to induce and generate tyrosine hydroxylase-positive dopaminergic neurons [32].

Ikeda et al. reported the directed differentiation of Rx- (an early bona fide marker of the eye field) and Pax6-positive retinal precursors from mouse ESCs under serum-free suspension conditions (SFEB culture) in the presence of Wnt and Nodal antagonists, Dkk1 and Lefty-A, respectively [33].

Osakada et al. defined a culture method that induces the development of both photoreceptors and RPE cells from mouse and human ESCs without the use of retinal tissues [5]. They also reported that the induced RPE cells functionally and morphologically mimicked natural RPE cells: they are hexagonal and pigmented, showing the presence of tight junctions, adhesive junctions, microvilli, and pigment granules [34].

Moreover, Haruta et al. were the first to report the transplantation of primate ESC-RPE into the subretinal space of the Royal College of Surgeons rats—a model for retinal degeneration—and showed that the RPE enhanced the survival of host photoreceptors [7]. These results provided a proof of concept that transplantation of pluripotent stem cell-derived RPE is a viable treatment option for retinal degeneration. Subsequent to this report, others have described other methods for differentiation and animal models [35–37]; however, ethical concerns regarding the use of human embryos and the potential of tissue rejection after transplantation prevented the use of ESCs for clinical applications.

8.3.4 iPSC as a Source for RPE

iPSCs, generated from reprogrammed adult somatic cells using the Yamanaka pluripotent transcription factors [2], have offered exciting new possibilities of producing patient-specific stem cell lines without the use of human embryonic cells. Retinal cells can be prepared from iPSCs based on the ESC culture method. Hirami et al. were the first to successfully produce retinal cells from mouse and human iPSCs by differentiating in vitro using a serum- and feeder-free method to avoid the risk of adverse immunologic responses and potential exposure to xenopathogens [6]. Jin et al. reported the production of intrinsic patient-specific iPSCs from patients with distinct mutations in the *RP1*, *RP9*, *PRPH2*, or *RHO* genes that recapitulated the disease feature in vitro [38]. These disease modeling techniques have contributed greatly to the elucidation of the retinal degeneration pathology.

We subsequently investigated the morphological and functional features of iPSC-RPE for its clinical adaptation. Kamao et al. assessed the quality, quantity, consistency, and safety of clinical-grade hiPSC-RPE sheets [39] and demonstrated that iPSC-RPE formed tight junctions, secreted growth factors, and showed phagocytotic ability and gene expression patterns akin to those of native RPE. Maeda et al. showed that iPSC-RPE possesses functional visual cycle enzymes in vitro and in vivo. Moreover, they reported that transplantation of iPSC-RPE into *Lrat−/−* and *Rpe65−/−* mice rescued vision in these animals [40]. These RPE cells expressed the typical RPE marker and showed secretory ability and physiological activity similar to those of native RPE in vivo and, thus, can be readily used as cells for transplantation.

8.3.5 Superiority of iPSC-RPE for Transplantation

The iPSC-RPE transplantation using human iPSCs was fast-tracked for its clinical use in patients with AMD for the following reasons: (1) the cellular functions and reproducibility of iPSC-RPE have been well described, as mentioned above; (2) the number of cells required for treatment is relatively small; (3) iPSC-RPE shows unique pigmentation during differentiation, which is useful for the identification, purification, and evaluation of these cells [39]; (4) clinical studies of human ESC-derived RPE for dry AMD had already been performed [41, 42]; and (5) RPE is an especially safe cell type and rarely forms tumors.

The safety and tumorigenicity of iPSC products are especially a great concern. Tumor formation from residual undifferentiated or incompletely differentiated human PSC-RPE is an issue that must be carefully evaluated. In 2013 and 2014, the Kawamata group reported a highly sensitive method to detect residual human iPSC based on qRT-PCR [43] and confirmed that iPSC-RPE possessed negligible tumorigenic potential, as shown by the results of subcutaneous tumorigenicity test using Matrigel and subretinal tumorigenicity test in NOG mice and nude rats, respectively [44]. The same group also reported that PEDF, which is secreted from primary or iPSC-RPE, induced apoptosis in iPSC, indicating that the tumor-forming potential of iPSCs can be suppressed by a simultaneous transplantation of iPSC-RPE [45]. Moreover, the inside of the eye abounds with retinoic acid, which has antitumor activity, and there has been no report of RPE tumor growth in the eye. Taken together, these findings suggested that RPE is a particularly safe cell type that seldom forms tumors, especially under a nonpermissive environment for tumor formation.

8.3.6 Production of iPSC-RPE for Its Clinical Application

The production of iPSC-RPE for clinical application requires strict quality control and accordance with good manufacturing practice (GMP). The GMP ministerial ordinance, issued by the Japanese government, stipulates the maintenance of both product safety and manufacturing.

In the production of RPE sheets for our clinical study, we confirmed that all culture reagents were traceable and secured, and a clinically usable manufacturing practice was adopted to ensure that the differentiation-induction process was suitable for a clinical use. For the iPSC culture, we utilized a cell processing facility with a GMP-compliant cell regulation room. All procedures were checked by personnel from three divisions, (1) the cell culture department, (2) the quality control department, and (3) the facility management department, and all processes including cell culture, quality control, and facility management were recorded in standard operation procedure (SOP) documents [46].

We generated two lines of patient-specific iPSCs using non-integrating episomal vectors and differentiated these cells into RPE cells as described previously. These iPSC-derived RPE cells showed DNA methylation and gene expression profiles consistent with those of RPE tissues. In addition, using single-cell qPCR, we detected levels of RPE-specific gene expression consistent with those of RPE tissues [8]. These cells were subsequently assessed for quality and safety before used for transplantation.

In whole genome sequencing, single nucleotide mutations were detected, but these were not identified as cancer driver mutations. Large de novo insertions or deletions or changes in DNA copy number were not detected in the protein-coding region. Furthermore, we confirmed that the plasmid DNA was not integrated into the genomic DNA [8].

As described above, many aspects of the cell safety have been examined. Nevertheless, this was the first clinical study of human iPSC-derived cells and has received great interest because of its long-term safety, such as the lack of tumor formation, cell hyperproliferation, and ectopic tissue formation after transplantation.

8.3.7 First Clinical Application of iPSC for AMD

The first clinical study of iPSC-RPE for neovascular AMD was initiated in August 2013; the patient was enrolled in November 2013 and received the iPSC-RPE sheet transplant for neovascular AMD in September 2014. The patient was a 77-year-old Japanese woman diagnosed with a polypoidal choroidal vasculopathy (a subtype of neovascular AMD). She received repeated intraocular injections of an anti-VEGF drug before the surgery and presented a gradual decrease in visual acuity for 5 years.

The patient underwent a surgery to remove the neovascular membrane and receive transplantation of the autologous iPSC-derived RPE cell sheet under the macular area (the center of visual function), using a surgical device consisted of a custom-designed hand piece and a cannula [47]. The surgery was successful and no major bleeding or other serious adverse events were observed. After the surgery, the graft sheet initially curled on its margin but gradually flattened by 8 weeks post-surgery. At 2 years following the surgery, the transplanted sheet remained intact as detected by both fundus photography and optical coherence tomography (OCT), and the graft area gradually expanded until approximately 2 years post-surgery; these findings indicated a successful engraftment without rejection [8].

A previous report showed that autologous iPSC-derived cells can cause immune-mediated rejection in mice [48]; however, the patient in our study did not show any sign of tissue rejection, even without the use of immunosuppressant drugs. Nevertheless, we advise caution in extrapolating the findings for other types of iPSC-based transplantation, especially because iPSC-derived RPE cells have been shown to inhibit T-cell activation [49], and the immunoprivileged nature of the sub-retinal space [50].

For the patient in our study, we did not observe improvement in postoperative best corrected visual acuity, which was maintained at approximately 0.1 (equivalent to 20/200 on the Snellen chart) throughout the follow-up period. With the enrolment of more patients, we should be able to conclusively determine whether iPSC-RPE can slow the progression of retinal degeneration or improve vision by restoring retinal functions. Thus far, we have not observed tumor formation; however, the long-term safety of this treatment must be assessed. We should emphasize that findings from a single case cannot fully clarify the risk or benefit associated with the procedure.

8.3.8 Allotransplantation Using iPSC-RPE

Autologous iPSC, which has lower chance for tissue rejection, is an ideal cell source for a transplant therapy; however, production of autologous iPSC lines is costly and time-consuming. To address this issue, we are currently exploring the use of allogenic RPE cells from iPSCs of HLA homozygote donors. The host immune response to the transplanted cells must be thoroughly examined for allogeneic grafts, and the importance of HLA matching in transplantation therapy has been demonstrated in hematopoietic stem cell transplantation [51].

We first investigated the presence of immune responses in in vitro and in vivo HLA-matched models. In vitro, iPSC-derived RPE cells express HLA class I/II antigens, but T cells do not respond to HLA-A-, HLA-B-, and HLA-DRB1-matched iPSC-derived RPE cells from HLA homozygote donors [49]. Moreover, in a mammalian in vivo model, no sign of tissue rejection was observed in MHC-matched iPSC-derived RPE allografts in the absence of immunosuppression. In contrast, immune response was detected around the graft and retinal tissue damage in the MHC-mismatched model [52]. Based on these findings, we hypothesize that iPSC-RPE from MHC homozygous donors can be used to treat retinal diseases in histocompatible recipients.

The Center for iPSC Research and Application (CiRA) of Kyoto University has provided guidance for our study, including to access the Japanese bone marrow bank, to create iPSCs with common HLA in homozygous Japanese population. The cells generated thus far matched the HLA genotypes of approximately 17% of the Japanese population [53]. We also investigated the frequency of HLA class I and II alleles and haplotypes among AMD patients in Japan and found that the frequency was not different than that of the Japanese population (manuscript submitted). Using this resource, we are currently performing larger, more efficient, and more informative clinical trials. This treatment strategy requires different forms of donor cells depending on the type and stage of the target disease, in a regenerative manner that is distinct from the present strategy using small molecule drugs.

8.4 ESC/iPSC-Photoreceptor Replacement Therapy

8.4.1 Target Disease for Photoreceptor Replacement Therapy

Retinitis pigmentosa (RP) is an inherited disease characterized by night blindness and progressive visual field loss starting from the mid-peripheral region in its early stage, which then progresses into a gradual loss of the peripheral visual field, development of tunnel vision, and ultimately severe visual impairment. Currently, treatment options for RP to improve visual function are limited to artificial retina [54, 55] and gene therapy [56] (to a limited state). RP is caused by the death of retinal photoreceptors; however, neural circuits from secondary neurons, such as bipolar cells, to the brain are thought to be preserved in progressive conditions [57]. It is thought that visual function may be restored by photoreceptor transplantation to reconstruct the neural network.

8.4.2 Treatment Concept of Retina or Photoreceptor Transplantation

Photoreceptor cells are the first-order neuron of the retina and are irreversibly lost in many retinal diseases; thus, goals of retinal regenerative medicine include photoreceptor transplantation and sufficient improvement of visual acuity.

Visual function can be considered restored if synapses are formed between photoreceptor cells and secondary neurons such as bipolar cells. However, the formation of this functional synaptic connection between photoreceptors and the inner retinal neural element is a major hurdle in regenerative medicine of photoreceptors.

From the 1980s, numerous studies have reported the transplantation of micro aggregates of retinal progenitor cells or retinal sheets in various models of retinal degeneration [14, 58–60]. However, these attempts had poor survival rate and synapse formation with secondary neurons; thus, satisfactory outcomes could not be obtained.

More recently, Radtke et al. reported that transplantation of fetal retinal tissue along with its RPE improved vision in patients with retinal degeneration [61]. Pearson et al. demonstrated the integration of postmitotic photoreceptor precursors into host retinas following transplantation, which retained the outer nuclear layer (ONL), differentiated into rod photoreceptors, and gained rod functional recovery [62]. These findings can serve as a proof of concept for therapeutic photoreceptor transplantation. However, Pearson et al. and others recently reinterpreted the findings and concluded that the repair following photoreceptor implantation was likely the result of cytoplasmic transfer from transplanted cells rather than due to direct integration of graft cells [63, 64]. Thus, whether transplanted photoreceptors can form a synapse with adult host retinal cells is yet to be determined.

8.4.3 Photoreceptors Generated from Stem Cells

Protocols to differentiate retinal tissue from human ESCs or iPSCs have been presented in several reports [33, 35]. Osakada et al. were able to induce Rx- or Mitf-positive retinal progenitor cells from feeders- and serum-free hESC [5] and successfully differentiated Rx- or Mitf-positive retinal progenitor cells into photoreceptors by subsequently treating with retinoic acid and taurine [34]. More recently, Eiraku et al. reported a groundbreaking study in which they created a self-organizing optic cup using three-dimensional culture of ESCs [65]. Using this method, retina-specific markers can be observed throughout the course of retinal development [66], allowing the preparation of retinal tissue at any developmental stage. Moreover, retinal tissue can be prepared as both cell suspension and sheets [67]. We previously observed that retinal cells may not survive long term in advanced retinal degeneration [68]; nevertheless, the aforementioned method for self-formation of three-dimensional neural retina tissues should allow the generation of sufficient quality, quantity, and purity of cells as the source for cell-based therapies for retinal degenerative diseases.

8.4.4 Treatment Proof of Concept

We believe that the first clinical application of photoreceptor replacement therapy should target cases of RP with complete depletion of photoreceptors or ONLs. Therefore, we have utilized an end-stage retina model to explore the viability of photoreceptor transplantation.

Assawachananont et al. reported the transplantation efficacy of murine ESC- or iPSC-derived retinal tissue into the retina of mice with progressive retinal degeneration [69]. After mESC/iPSC transplantation, the retinas differentiated into structured ONL with mature outer segments. Additionally, immunohistochemical analysis showed synaptic connections in a direct-contact pattern between host bipolar cells and the graft in the host-graft interface. We have also developed a monkey model of retinal degeneration. Shirai et al. reported monkey models of retinal degeneration using a subretinal injection of cobalt chloride or laser photocoagulation and confirmed the structured graft maturation and potential formation of synaptic contacts between the graft and host cells, which were previously shown using similar primate models [70]. However, these findings cannot conclusively show whether an ESC-/iPSC-derived structured, retina-like sheet can restore visual functions.

We recently demonstrated a direct contact between the host bipolar cell-end and the presynaptic terminal of the graft photoreceptor, as visualized by gene labeling, by directly implanting retinal tissues derived from murine iPSC (miPSC retina) into a mouse model of end-stage retinal degeneration (rd1). We also showed photo-responsive behavior in transplanted rd1 mice and recorded responses from transplanted host retinas by ex vivo micro-electroretinography and ganglion cell recordings using a multiple-electrode array system [71]. We believe that our findings established a new proof of concept for transplanting ESC/iPSC retinas to restore vision in patients with end-stage retinal degeneration.

8.4.5 Environmental Factors Supporting Retinal Graft Survival

Adequate conditions for graft cell survival in photoreceptor transplantation, such as the time window during the course of retinal degeneration and environmental factors that are permissive for retinal graft cell survival, should also be thoroughly assessed. We investigated the retinal degeneration process of rd1 mouse and observed the infiltration of microglia that are positive for ionized calcium-binding adapter molecule 1 (Iba1). The expression of glial fibrillary acidic protein (GFAP) was increased at the peak of rod death but was reduced thereafter until gliosis began to take place, in which enhanced GFAP expression was again observed [68]. These findings indicated that photoreceptors should be transplanted within a finite time window to avoid microglia activation by inflammation and barrier formation by glial cells. We expect that transplantation efficiency is improved by modifying the host environment with drugs that promote synapse formation or eliminate glial scarring. We are currently preparing the application of photoreceptor replacement therapy in patients by assessing suitable transplantation conditions.

8.5 Conclusion

Regeneration of the retina, which is part of the cranial nerve, is a challenging task. Nevertheless, sufficient advances had been made through efforts by our group and others, which enabled us to conduct the first clinical trial using iPSCs for the retina.

In this chapter, we focused mainly on background studies we have performed thus far; however, achieving visual reconstruction using stem cells is still a challenge for many researchers. In Japan, iPSCs have received great attention, and expectation for regenerative medicine is quite high. We should note that the use of iPSCs as a treatment strategy is still in its infancy, and the therapeutic effect has been limited. We believe that a comprehensive strategy that incorporates both regenerative medicine and rehabilitation is crucial; thus, our future efforts will focus on maximizing the therapeutic effect by employing a comprehensive medical treatment that includes low-vision care.

References

1. Thomson JA, Itskovitz-Eldor J, Shapiro SS, et al. Embryonic stem cell lines derived from human blastocysts. Science. 1998;282(5391):1145–7.
2. Takahashi K, Yamanaka S. Induction of pluripotent stem cells from mouse embryonic and adult fibroblast cultures by defined factors. Cell. 2006;126(4):663–76.
3. Takahashi K, Tanabe K, Ohnuki M, et al. Induction of pluripotent stem cells from adult human fibroblasts by defined factors. Cell. 2007;131(5):861–72.
4. Yamanaka S. A fresh look at iPS cells. Cell. 2009;137(1):13–7.
5. Osakada F, Ikeda H, Mandai M, et al. Toward the generation of rod and cone photoreceptors from mouse, monkey and human embryonic stem cells. Nat Biotechnol. 2008;26(2):215–24.
6. Hirami Y, Osakada F, Takahashi K, et al. Generation of retinal cells from mouse and human induced pluripotent stem cells. Neurosci Lett. 2009;458(3):126–31.
7. Haruta M, Sasai Y, Kawasaki H, et al. In vitro and in vivo characterization of pigment epithelial cells differentiated from primate embryonic stem cells. Invest Opthalmol Vis Sci. 2004;45(3):1020.
8. Mandai M, Watanabe A, Kurimoto Y, et al. Autologous induced stem-cell-derived retinal cells for macular degeneration. N Engl J Med. 2017;376(11):1038–46.
9. Price J, Turner D, Cepko C. Lineage analysis in the vertebrate nervous system by retrovirus-mediated gene transfer (retrovirus vectors!/-galactosidase/neural progenitors/retina). Proc Natl Acad Sci U S A. 1987;84:156–60.
10. Luskin MB. Restricted proliferation and migration of postnatally generated neurons derived from the forebrain subventricular zone. Neuron. 1993;11(1):173–89.
11. Kuhn HG, Dickinson-Anson H, Gage FH. Neurogenesis in the dentate gyrus of the adult rat: age-related decrease of neuronal progenitor proliferation. J Neurosci. 1996;16(6):2027–33.
12. Vescovi A, Reynolds BA, Fraser DD, Weiss S. bFCF regulates the proliferative fate of unipotent (neuronal) and bipotent (neuronallastroglial) EGF-generated CNS progenitor cells. Neuron. 1993;11:951–66.
13. Gage FH. The adult rat hippocampus contains primordial neural stem cells.pdf. Mol Cell Neurosci. 1997;8:389–404.
14. Takahashi M, Palmer TD, Takahashi J, Gage FH. Widespread integration and survival of adult-derived neural progenitor cells in the developing optic retina. Mol Cell Neurosci. 1998;12:340–8.
15. Akagi T, Haruta M, Akita J, et al. Different characteristics of rat retinal progenitor cells from different culture periods. Neurosci Lett. 2003;341(3):213–6.
16. Bok D. The retinal pigment epithelium: a versatile partner in vision. J Cell Sci Suppl. 1993;17:189–95.
17. Strauss O. The retinal pigment epithelium in visual function. Physiol Rev. 2005;85(3):845–81.

18. Adamis AP, Shima DT, Yeo KT, et al. Synthesis and secretion of vascular permeability factor/vascular endothelial growth factor by human retinal pigment epithelial cells. Biochem Biophys Res Commun. 1993;193(2):631–8.
19. Dawson DW, Volpert OV, Gillis P, et al. Pigment epithelium-derived factor: a potent inhibitor of angiogenesis. Science. 1999;285(5425):245–8.
20. Khaliq A, Patel B, Jarvis-Evans J, et al. Oxygen modulates production of bFGF and TGF-beta by retinal cells in vitro. Exp Eye Res. 1995;60(4):415–23.
21. Friedman DS, O'Colmain BJ, Munoz B, et al. Prevalence of age-related macular degeneration in the United States. Arch Ophthalmol. 2004;122(4):564–72.
22. Gass JD. Biomicroscopic and histopathologic considerations regarding the feasibility of surgical excision of subfoveal neovascular membranes. Am J Ophthalmol. 1994;118(3):285–98.
23. Yasuda M, Kiyohara Y, Hata Y, et al. Nine-year incidence and risk factors for age-related macular degeneration in a defined Japanese population the Hisayama study. Ophthalmology. 2009;116(11):2135–40.
24. Brown DM, Michels M, Kaiser PK, et al. Ranibizumab versus verteporfin photodynamic therapy for neovascular age-related macular degeneration: two-year results of the ANCHOR study. Ophthalmology. 2009;116(1):57–65.e5.
25. Martin DF, Maguire MG, Fine SL, et al. Ranibizumab and bevacizumab for treatment of neovascular age-related macular degeneration: two-year results. Ophthalmology. 2012;119(7):1388–98.
26. Rivera A, Fisher SA, Fritsche LG, et al. Hypothetical LOC387715 is a second major susceptibility gene for age-related macular degeneration, contributing independently of complement factor H to disease risk. Hum Mol Genet. 2005;14(21):3227–36.
27. Yang Z, Camp NJ, Sun H, et al. A variant of the HTRA1 gene increases susceptibility to age-related macular degeneration. Science. 2006;314(5801):992–3.
28. Group A-REDSR. A randomized, placebo-controlled, clinical trial of high-dose supplementation with vitamins C and E, beta carotene, and zinc for age-related macular degeneration and vision loss: AREDS report no. 8. Arch Ophthalmol. 2001;119(10):1417–36.
29. Zarbin MA, Rosenfeld PJ. Pathway-based therapies for age-related macular degeneration: an integrated survey of emerging treatment alternatives. Retina. 2010;30(9):1350–67.
30. Algvere PV, Berglin L, Gouras P, Sheng Y. Transplantation of fetal retinal pigment epithelium in age-related macular degeneration with subfoveal neovascularization. Graefes Arch Clin Exp Ophthalmol. 1994;232(12):707–16.
31. van Zeeburg EJ, Maaijwee KJ, Missotten TO, et al. A free retinal pigment epithelium-choroid graft in patients with exudative age-related macular degeneration: results up to 7 years. Am J Ophthalmol. 2012;153(1):120–7.e2.
32. Kawasaki H, Suemori H, Mizuseki K, Watanabe K, Urano F, Ichinose H, Haruta M, Takahashi M, Yoshikawa K, Nishikawa S, Nakatsuji N, Sasai Y. Generation of dopaminergic neurons and pigmented epithelia from primate ES cells by stromal cell-derived inducing activity. Proc Natl Acad Sci U S A. 2002;99(3):1580–5.
33. Ikeda H, Osakada F, Watanabe K, Mizuseki K, Haraguchi T, Miyoshi H, Kamiya D, Honda Y, Sasai N, Yoshimura N, Takahashi M, Sasai Y. Generation of Rx Pax6neural retinal precursors from embryonic stem cells. Proc Natl Acad Sci U S A. 2005;102(32):11331–6.
34. Osakada F, Jin ZB, Hirami Y, et al. In vitro differentiation of retinal cells from human pluripotent stem cells by small-molecule induction. J Cell Sci. 2009;122(Pt 17):3169–79.
35. Lund RD, Wang S, Klimanskaya I, et al. Human embryonic stem cell-derived cells rescue visual function in dystrophic RCS rats. Cloning Stem Cells. 2006;8(3):189–99.
36. Meyer JS, Shearer RL, Capowski EE, et al. Modeling early retinal development with human embryonic and induced pluripotent stem cells. Proc Natl Acad Sci U S A. 2009;106(39):16698–703.
37. Amirpour N, Nasr-Esfahani MH, Esfandiari E, et al. Comparing three methods of co-culture of retinal pigment epithelium with progenitor cells derived human embryonic stem cells. Int J Prev Med. 2013;4(11):1243–50.

38. Jin ZB, Okamoto S, Osakada F, et al. Modeling retinal degeneration using patient-specific induced pluripotent stem cells. PLoS One. 2011;6(2):e17084.
39. Kamao H, Mandai M, Wakamiya S, Ishida J, Goto K, Ono T, Suda T, Takahashi M, Kiryu J. Objective evaluation of the degree of pigmentation in human induced pluripotent stem cell–derived RPE. Invest Ophthalmol Vis Sci. 2014;55:8309–18.
40. Maeda T, Lee MJ, Palczewska G, et al. Retinal pigmented epithelial cells obtained from human induced pluripotent stem cells possess functional visual cycle enzymes in vitro and in vivo. J Biol Chem. 2013;288(48):34484–93.
41. Schwartz SD, Hubschman J-P, Heilwell G, et al. Embryonic stem cell trials for macular degeneration: a preliminary report. Lancet. 2012;379(9817):713–20.
42. Schwartz SD, Regillo CD, Lam BL, et al. Human embryonic stem cell-derived retinal pigment epithelium in patients with age-related macular degeneration and Stargardt's macular dystrophy: follow-up of two open-label phase 1/2 studies. Lancet. 2015;385(9967):509–16.
43. Kuroda T, Yasuda S, Kusakawa S, et al. Highly sensitive in vitro methods for detection of residual undifferentiated cells in retinal pigment epithelial cells derived from human iPS cells. PLoS One. 2012;7(5):e37342.
44. Kanemura H, Go MJ, Shikamura M, et al. Tumorigenicity studies of induced pluripotent stem cell (iPSC)-derived retinal pigment epithelium (RPE) for the treatment of age-related macular degeneration. PLoS One. 2014;9(1):e85336.
45. Kanemura H, Go MJ, Nishishita N, et al. Pigment epithelium-derived factor secreted from retinal pigment epithelium facilitates apoptotic cell death of iPSC. Sci Rep. 2013;3:2334.
46. Takahashi M. Retinal cell therapy using iPS cells. Nippon Ganka Gakkai Zasshi. 2016;120(3):210–24; discussion 25
47. Kamao H, Mandai M, Ohashi W, et al. Evaluation of the surgical device and procedure for extracellular matrix-scaffold-supported human iPSC-derived retinal pigment epithelium cell sheet transplantation. Invest Ophthalmol Vis Sci. 2017;58(1):211–20.
48. Zhao T, Zhang ZN, Rong Z, Xu Y. Immunogenicity of induced pluripotent stem cells. Nature. 2011;474(7350):212–5.
49. Sugita S, Iwasaki Y, Makabe K, et al. Lack of T cell response to iPSC-derived retinal pigment epithelial cells from HLA homozygous donors. Stem Cell Rep. 2016;7(4):619–34.
50. Zamiri P, Masli S, Streilein JW, Taylor AW. Pigment epithelial growth factor suppresses inflammation by modulating macrophage activation. Invest Ophthalmol Vis Sci. 2006;47(9):3912–8.
51. Lee SJ, Klein J, Haagenson M, et al. High-resolution donor-recipient HLA matching contributes to the success of unrelated donor marrow transplantation. Blood. 2007;110(13):4576–83.
52. Sugita S, Iwasaki Y, Makabe K, et al. Successful transplantation of retinal pigment epithelial cells from MHC homozygote iPSCs in MHC-matched models. Stem Cell Rep. 2016;7(4):635–48.
53. Saito S, Ota S, Yamada E, Inoko H, Ota M. Allele frequencies and haplotypic associations defined by allelic DNA typing at HLA class I and class II loci in the Japanese population. Tissue Antigens. 2000;56:522–9.
54. Stronks HC, Dagnelie G. The functional performance of the Argus II retinal prosthesis. Expert Rev Med Devices. 2014;11(1):23–30.
55. da Cruz L, Dorn JD, Humayun MS, et al. Five-year safety and performance results from the Argus II retinal prosthesis system clinical trial. Ophthalmology. 2016;123(10):2248–54.
56. Cideciyan AV, Jacobson SG, Beltran WA, et al. Human retinal gene therapy for Leber congenital amaurosis shows advancing retinal degeneration despite enduring visual improvement. Proc Natl Acad Sci U S A. 2013;110(6):E517–25.
57. Jacobson SG, Matsui R, Sumaroka A, Cideciyan AV. Retinal structure measurements as inclusion criteria for stem cell-based therapies of retinal degenerations. Invest Ophthalmol Vis Sci. 2016;57(5):ORSFn1–9.
58. Young MJ, Ray J, Whiteley SJ, et al. Neuronal differentiation and morphological integration of hippocampal progenitor cells transplanted to the retina of immature and mature dystrophic rats. Mol Cell Neurosci. 2000;16(3):197–205.

59. Nishida A, Takahashi M, Tanihara H, et al. Incorporation and differentiation of hippocampus-derived neural stem cells transplanted in injured adult rat retina. Invest Ophthalmol Vis Sci. 2000;41(13):4268–74.
60. Ooto S, Akagi T, Kageyama R, Akita J, Mandai M, Honda Y, Takahashi M. Potential for neural regeneration after neurotoxic injury in the adult mammalian retina. Proc Natl Acad Sci U S A. 2004;101(37):13654–9.
61. Radtke ND, Aramant RB, Petry HM, et al. Vision improvement in retinal degeneration patients by implantation of retina together with retinal pigment epithelium. Am J Ophthalmol. 2008;146(2):172–82.
62. Pearson RA, Barber AC, Rizzi M, et al. Restoration of vision after transplantation of photoreceptors. Nature. 2012;485(7396):99–103.
63. Pearson RA, Gonzalez-Cordero A, West EL, et al. Donor and host photoreceptors engage in material transfer following transplantation of post-mitotic photoreceptor precursors. Nat Commun. 2016;7:13029.
64. Santos-Ferreira T, Llonch S, Borsch O, et al. Retinal transplantation of photoreceptors results in donor-host cytoplasmic exchange. Nat Commun. 2016;7:13028.
65. Eiraku M, Takata N, Ishibashi H, et al. Self-organizing optic-cup morphogenesis in three-dimensional culture. Nature. 2011;472(7341):51–6.
66. Nakano T, Ando S, Takata N, et al. Self-formation of optic cups and storable stratified neural retina from human ESCs. Cell Stem Cell. 2012;10(6):771–85.
67. Gonzalez-Cordero A, West EL, Pearson RA, et al. Photoreceptor precursors derived from three-dimensional embryonic stem cell cultures integrate and mature within adult degenerate retina. Nat Biotechnol. 2013;31(8):741–7.
68. Mandai M, Homma K, Okamoto S, et al. Adequate time window and environmental factors supporting retinal graft cell survival in rd mice. Cell Med. 2012;4(1):45–54.
69. Assawachananont J, Mandai M, Okamoto S, et al. Transplantation of embryonic and induced pluripotent stem cell-derived 3D retinal sheets into retinal degenerative mice. Stem Cell Rep. 2014;2(5):662–74.
70. Shirai H, Mandai M, Matsushita K, et al. Transplantation of human embryonic stem cell-derived retinal tissue in two primate models of retinal degeneration. Proc Natl Acad Sci U S A. 2016;113(1):E81–90.
71. Mandai M, Fujii M, Hashiguchi T, et al. iPSC-derived retina transplants improve vision in rd1 end-stage retinal-degeneration mice. Stem Cell Rep. 2017;8(1):69–83.

Chapter 9
Organoid Models of Development and Disease Towards Therapy

Yasunori Nio and Takanori Takebe

Abstract Historical developmental biology experiments demonstrated the remarkable capacity of reaggregated vertebrate cells for self-organization, now exploited to rebuild human tissues, reminiscent of native organs, from somatic or pluripotent stem cells, namely, an organoid. Organoid technology is thus rapidly evolving and becoming an independent research field due to its potential for modelling human development and disease. Coupled with patient-derived stem cells, diseased organoid recapitulates a pathological state in a dish, promoting personalized medicine and drug development. Ultimately, organoid transplantation paves a way for organ replacement strategies against end-stage diseases. This article summarizes the evolutionary organoid technology backed by developmental biology and outlines its phenomenal potential for future therapeutic applications.

Keywords Organoid · Three-dimensional cell culture model · Induced pluripotent stem cell · Embryonic stem cell · Organ development

---

Y. Nio
T-CiRA, Regenerative Medicine Unit, Takeda Pharmaceutical Company Limited, Fujisawa City, Kanagawa, Japan

T. Takebe (✉)
T-CiRA, Regenerative Medicine Unit, Takeda Pharmaceutical Company Limited, Fujisawa City, Kanagawa, Japan

Department of Regenerative Medicine, Yokohama City University Graduate School of Medicine, Yokohama, Kanagawa, Japan

Organogenesis and Neogenesis Unit, Institute of Research, Tokyo Medical and Dental University, Tokyo, Japan

Division of Gastroenterology, Hepatology & Nutrition, Developmental Biology, Cincinnati Children's Hospital Medical Center, Cincinnati, OH, USA

Center for Stem Cell and Organoid Medicine (CuSTOM), Cincinnati Children's Hospital Medical Center, Cincinnati, OH, USA

Institute of Research, Tokyo Medical and Dental University, Tokyo, Japan
e-mail: takanori.takebe@cchmc.org

H. Inoue, Y. Nakamura (eds.), *Medical Applications of iPS Cells*, Current Human Cell Research and Applications,
https://doi.org/10.1007/978-981-13-3672-0_9

9.1 Introduction

Recently, human organoids have received international attention as in vitro culture systems, where stem cells self-organize into three-dimensional structures reminiscent of developing human organs. For example, due to their potential for mirroring the human phenotype in a dish, organoids can serve as an avatar for patients' pathobiology and individual responses to potential therapy, offering a new approach towards drug development and precision medicine. Ultimately, organoids, coupled with an evolving tissue engineering approach, will provide an organ replacement strategy to end-stage disease conditions when transplanted. However, such exciting implications for personalized and regenerative medicine, hereafter defined as "organoid medicine", remain challenging.

Herein, we comprehensively review the development history of 3D organoid models and their potential applications for drug discovery and transplant applications to provide effective cure for patients with a goal of accelerating the community efforts to realize organoid medicine.

9.1.1 Historical Developments in Tissue Organoids (3D Models for Organ Development)

The evolution of the organoid has been driven by the number of scientists with key discovery elements that is far beyond describable. Therefore, for the simplicity reasons, we herein summarize organoid development history from three major angles (Fig. 9.1).

9.1.1.1 First (Organoid 1.0): Reaggregation and Self-Organization

The first description of key principles in organoid biology dated back in the early 1900s. Wilson et al. conducted ground-breaking reaggregation experiments using sponges [1]. This investigation enhanced self-organization experiments by using isolated cells from primary tissues. Based on a proposal by Steinberg et al. [2], differential adhesion hypothesis first defined cell sorting as a primary mechanism of self-organization. Later, Takeichi et al. showed that cadherin-mediated processes are necessary for self-organization [3].

9.1.1.2 Second (Organoid 2.0): Laminin-Rich Basement Membrane Matrix (BM)

Bissel et al. applied laminin-rich BMs extracted from Engelbreth-Holm-Swarm (EHS) mice carcinoma into mammary epithelial cell culture, enabling the organoid generation through the cell polarity induction after gel implantation [4]. Similar proof-of-principle is confirmed in the generation of lung and kidney organoid

Fig. 9.1 Evolutionary history of organoid technology

formation [5, 6]. Currently, most organoid-related research involves the use of carcinoma-derived BM to form organoids, hindering the clinical applications of organoids due to potential tumorigenicity. Recent seminal development of designer BMs will alleviate this concern towards organoid transplant applications in the future [7, 8].

9.1.1.3 Third (Organoid 3.0): Human Stem Cells

Given the significant advances that have taken place in stem cell biology and cell fate reprogramming, researchers now have the ability to manipulate various stem cell fates. Most notably, two pluripotent stem cells become available for the study of organoid technology: embryonic stem (ES) cell [9] as well as induced pluripotent stem (iPS) cells [10]. For example, Eiraku et al. investigated brain cortex organoids [11]. Organoid generation is further accelerated with the use of somatic stem cells, as shown in gut organoids [12]. Remarkably, in 2011, Eiraku et al. asked an elaborated developmental biology question with the advantage of organoid system, highlighting the potential utility of optic cup organoids for studying mechanistic basis of morphogenesis [13]. Recently, gut [14], liver [15] and brain [16], and kidney organoids [17, 18] were also reported from healthy and diseased pluripotent stem cells. Given most these organoids can be derived from human stem cells, human organoids have thus emerged as a powerful model to study biology of human in a dish.

Now, we are approaching the next generation of the organoid revolutionary process, namely, an Organoid 4.0 era. We argue that efforts will be shifting, rather just generating organoids, to elucidate the scientific and medical utility of organoid as a tool for applications through active integration of academic research, clinical care, and industrial practices, which in other words is organoid medicine.

9.2 Tissue Organoids

In this section, we provide a review of state-of-the-art organoid technologies in diverse tissues (Fig. 9.2).

9.2.1 Liver Organoids

A self-organizing tissue-based approach, coupled with iPSC technology, is expected to achieve valuable outcomes in regenerative medicine and drug development. As a conventional strategy, two-dimensional (2D) and unidirectional differentiation platforms have enabled dramatic advances in the ability to control the specific fate of iPSCs [19–21], although it has been difficult to establish models of complex organ functions owing to tissue immaturity or limited functionality [22]. Recent

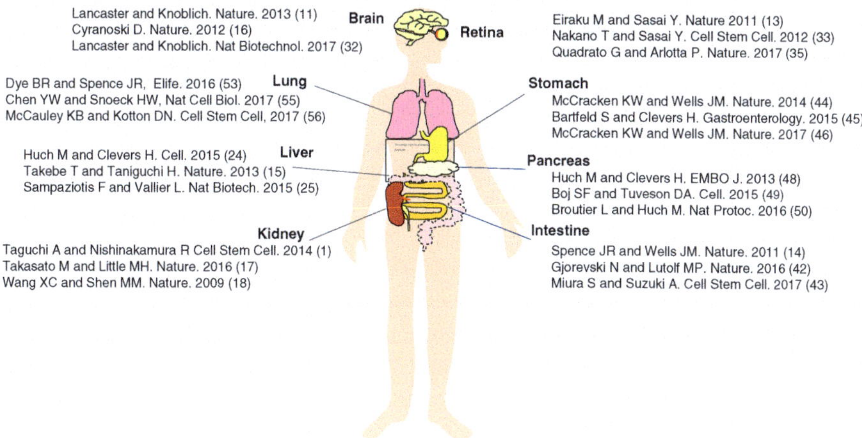

Fig. 9.2 Pluripotent stem cell (PSC)-derived organoids. State-of-the-art publications of developing diverse organoids from human embryonic stem cell or induced pluripotent stem cells

demonstration of hepatic organoid generation from human biopsy sample is promising [23, 24], especially revolutionizing personalized medicine application-based organoids from patients with diverse forms of liver cancers. The other exciting implication is to generate cholangicoytic organoids from human iPSC [25]. For building additional complexity, it is essential to incorporate multiple cell lineages, including developing vessels. Using a self-condensation method, recently, self-organizing vascularized organoids were generated in diverse systems together with human mesenchymal and endothelial progenitors [26]. In addition, iPSC-derived liver organoids were shown to anastomose with host vessels in vivo within 48 h and provide functional rescue against liver failure [15], highlighting potential applications in regenerative medicine, developmental biology, and drug discovery.

9.2.2 *Brain Organoids*

Recently, neural tube-like structures called neural rosettes were established from isolated neuroepithelial cells and from the directed differentiation of iPSCs [11, 27–29]. Because neural rosettes recapitulate apical-basal polarity and exhibit spontaneous radial organization similar to that of the neural tube, they are more capable of recapitulating many aspects of brain development. However, it has many limitations in modelling the overall organization of the developing brain. Therefore, alternative 3D culture methods—with the potential to recapitulate brain tissue organization—have been used extensively for investigations in recent years. In particular, Sasai et al. focused on developing 3D models of various isolated brain regions from mouse or human iPSCs [30]. Lancaster and Knoblich further advanced this approach by generating cerebral organoids, which are single neural organoids that have the features of several different brain regions [16]. Like the Sasai method, this approach starts with floating embryoid bodies (EBs), but growth factors are not added to drive particular brain region identities. Instead, the aggregates are embedded in a laminin-rich BM secreted by the Engelbreth-Holm-Swarm tumour cell line. These BMs promote proliferation of large neuroepithelial buds which can develop into various brain regions. Cerebral organoids can grow up to several millimetres using a spinning bioreactor, which provides nutrient and oxygen exchange. This expansion allows the formation of a variety of brain regions, including the dorsal cortex, ventral forebrain, midbrain-hindbrain boundary, choroid plexus, and hippocampus [16]. Moreover, to enhance neuroectoderm formation and improved cortical development, they combined these cerebral organoid techniques with poly(lactide-co-glycolide) copolymer (PLGA) fibre microfilaments as a floating scaffold to generate elongated embryoid bodies. Microfilament-engineered cerebral organoids (enCORs) enhanced neuroectoderm formation and improved cortical development. Therefore, this enCOR model allows for the study of neuronal migration with distinctive radial organization of the cerebral cortex. Combination study between bioengineering and self-organization would allow for reliable forebrain formation without sacrificing tissue complexity in brain organoid [31].

9.2.3 Retinal Organoids

The evolution of iPSC-derived retinal organoids was built upon a foundation rooted in developmental biology. Embryoid bodies are grown in minimal medium to generate neuroectoderm [13, 27]. Matrigel is dissolved in the medium at an early stage to allow the formation of more rigid neuroepithelial tissues, an essential step for retinal epithelium formation. This promotes the formation of retinal primordial tissue buds similar to the optic vesicle. The resulting optic cup organoids very closely mimic early retinas. They display proper markers of neural retinas and retinal pigmented epithelium; additionally, they exhibit retinal stratification with proper apical-basal polarity and undergo morphological tissue shape changes that mimic the stepwise evagination and invagination of the optic cup in vivo. These human retinal organoids show a number of human-specific features as well and exhibit certain tissue morphological differences, such as apical nuclear positioning [32]. Zhong et al. reported that hiPSC-derived 3D retinal cups could recapitulate spatiotemporally each of the main steps of retinal development observed in vivo and contained all major retinal cell types arranged in their proper layers. This hiPSC-derived retinal organoid achieves advanced maturation, showing the beginning of outer-segment-disc formation and photosensitivity [33]. Recently, Quadrato et al. showed that cell diversity and network dynamics were controlled by light stimulation of photoreceptor-like cells in human brain organoids. This phenomenon indicated that brain organoids establish spontaneously active neuronal networks and generate functional photosensitive cells, which may in the future permit modulation of network activity using physiological sensory mechanisms [34].

9.2.4 Kidney Organoids

Like many of the tissues for which organoids have been developed, evidence that kidney tissues may be capable of self-organization comes from early reaggregation experiments in chick embryonic kidneys [35]. The resultant self-developed tissues displayed nephron components including the collecting duct, distal and proximal tubules, and glomeruli with allantoic vessels after transplantation of the chick allantoic membrane. Recently, a renal precursor tissue derived from the intermediate mesoderm, or the ureteric epithelium, can be generated from human iPSCs via a mesodermal specification step [36–38]. When mouse embryonic kidneys are reaggregated after dissociation, they are self-organized into 3D ureteric bud structures. The metanephric mesenchyme can be derived from human and mouse embryoid bodies through sequential exposure to defined soluble factors. When metanephric mesenchyme and spinal cord tissue, a nephric inducer, are cocultured, nephric tubules and nascent glomeruli are well organized [17]. Kidney organoids have been shown to contain over 500 nephrons with defined glomeruli comprising a Bowman's capsule with podocytes and connect to proximal tubules. Although these organoids are more likely to mimic an embryonic kidney rather than an adult kidney, they contained a mixture of nephron progenitors and formed urine-collecting ducts. Musah et al. recently established human podocytes that expressed podocyte markers such as nephrin, WT1, and podocin [39]. As shown above, each parts of kidney organoids are getting established, and interconnectivity or more contagious structure establishment would be a real challenge.

9.2.5 Intestinal Organoids

The gastrointestinal (GI) tract develops primarily from the endoderm [27], which forms an epithelial tube that develops into three distinct portions: the foregut, midgut, and hindgut [40]. The hindgut gives rise to the remaining portion of the colon, or large intestine, and the rectum. This knowledge of Wnts and FGFs provides the foundation on which human intestinal organoids are built [14]. Clevers et al. showed that adult intestinal stem cells could form organoids when subjected to 3D Matrigel culture [12]. Similarly, hindgut spheroids generated from human iPSCs can be grown under Matrigel 3D conditions and will further develop to mature intestinal organoids [14]. Gjorevski et al. showed a fully defined synthetic hydrogel that mimics the extracellular matrix to support in vitro growth of intestinal stem cells and organoids. The hydrogel allows sophisticated control over the chemical and physical in vitro niche and enables identification of regulatory properties of the matrix [41]. Identification of organoid-forming intestinal progenitor cells by direct lineage reprogramming via cdx2, foxa3, hnf4, and gata6 are new potential approach to obtain cell source for generating intestinal organoids [42]. Intestinal organoids develop crypt-villus structures with stratified epithelium containing all the main cell lineages of the gut [27].

9.2.6 Stomach Organoids

The stomach is derived from the posterior foregut. McCracken and Wells et al. developed a protocol for generating stomach organoids from human PSC. They used activin treatment in human iPSCs to generate definitive endoderm, followed by the addition of BMP inhibitors and FGF and Wnt activators towards a foregut fate [43]. When retinoic acid was added, the organoids were specified to have a posterior foregut fate. High concentrations of epidermal growth factor (EGF) changed these cells into human gastric organoids, which progressed through molecular and morphogenic stages that resembled those of the developing antrum of the mouse stomach. The organoids included primitive gastric gland- and pit-like domains, proliferative zones with Lgr5-positive stem cells, mucous cells, and a host of gastric endocrine cells. Bartfeld et al. reported that human stomach organoids still maintain many characteristics of the original tissue after long-term culturing [36, 44]. Recently, McCracken et al. investigated Wnt/β-catenin promotes gastric fundus specification in mice and humans. Disruption of Wnt/β-catenin signalling in mouse embryos led to conversion of fundic to antral epithelium, while β-catenin activation in hPSC-derived foregut progenitors promoted the development of human fundic-type gastric organoids (hFGOs). These hFGOs could be a powerful new model for studying the development of the human fundus and the molecular basis of human gastric physiology, pathophysiology, and drug discovery [45].

9.2.7 Pancreas Organoids

During embryonic development, the pancreas develops in both the ventral and dorsal anterior foregut. The embryonic pancreas progenitors grow into 3D organoids which can be expanded for up to 2 weeks in culture and retain their potential to differentiate into acinar, ductal, or endocrine lineages [46]. The embryonic pancreas organoids show rapid proliferation and can recapitulate the branching structure of the pancreatic epithelium. Because the cells maintain their ability to efficiently differentiate into all three lineages, the system is useful for studying signals which bias cell fate decisions towards particular lineages. Recently the long-term culture of organoids derived from the adult mouse [47] and human [48] pancreas were established. Huch et al. defined culture conditions which promoted the growth of organoids by adding Rspondin-1 and Lgr5 ligand. In these conditions, mouse pancreas organoids could be expanded in culture for at least 5 months. For pancreas organoid culture, not HGF but Noggin was an essential growth factor [47]. So far, the direct differentiation of these cells along endocrine lineages in vitro has proven unsuccessful. However, when adult pancreas organoid cells were mixed with embryonic E13 mouse pancreas cells and immediately transplanted into the kidney capsule of immunocompromised mice, the adult-derived organoid cells readily differentiated into fully mature hormone excretion cells such as insulin, glucagon, and somatostatin-positive cells in vivo [47]. Although using similar culture conditions, human pancreas duct cells can be expanded in vitro for 4–5 months; endocrine secretion potential has not been analysed yet [48, 49]. Functional test such as insulin release would be undertaken in the near future.

9.2.8 Lung Organoids

Lung organoids have been grown from the embryonic lungs [50], mouse and human adult lungs [51], and human PSC and iPSC [52–55]. In mouse, lung development begins after establishment of the primary germ layers with two primary endoderm buds being visible at about E9.5 [56]. Although E12.5 mouse lung can be cultured in vitro on an airway interface in 2D where they will both grow and differentiate [57], comparing 3D culture, their size, cellular complexity, and growth in 2D plane was unsuitable for many experiments [58]. Mixed cell populations obtained from dissociated whole E17.5 mouse lungs have been grown in 3D on various matrics [50]. Such cultures self-organize into spheroids containing branched epithelial structures surrounded by mesenchyme, grow, and show signs of alveolar and bronchiolar differentiation [59, 60]. For disease modelling, generating mature disease-specific human lung cells is very important because murine models often do not show complete phenotype of human lung disease. Although human pluripotent stem cell (hPSC)-derived tissues often remain developmentally immature in vitro, they become more adult-like in their structure, cellular diversity, and function following transplantation into immunocompromised mice resulting in airway-like structures that were remarkably similar to the native adult human lung [52]. Using single-cell RNA sequencing, Kim et al. investigated that Lgr6+ mesenchymal cells in the adult lung promote airway differentiation of epithelial progenitors via Wnt-Fgf10 cooperation and Lgr5+ cells are located in alveolar compartments and are sufficient to promote alveolar differentiation of epithelial progenitors through Wnt activation. This identification of region- and lineage-specific crosstalk between epithelium and their neighbouring mesenchymal partners provides new understanding of how different cell types are maintained in the adult lung [53]. Regarding iPSC, Hans et al. reported the generation from hPSCs of lung bud organoids (LBOs) that contain mesoderm and pulmonary endoderm and develop into branching airway and early alveolar structures after xenotransplantation and in Matrigel 3D culture. Using these methods and induction of mutation in in Hermansky-Pudlak syndrome (HPS)-1 gene, they could recapitulate fibrotic lung disease in vitro which causes an early-onset form of intractable pulmonary fibrosis [54]. Kotton et al. showed that cyclical modulation of the canonical Wnt signalling pathway enables rapid directed differentiation of human iPSCs via an NKX2-1+ progenitor intermediate into functional proximal airway organoids. They also could generate cystic fibrosis patient-specific iPSC-derived airway organoids with a defect in forskolin-induced swelling, and their approach has many potential applications in modelling and drug screening for airway diseases [55].

9.3 3D Organoids for Disease Treatment

9.3.1 Introduction

Owing to the resemblance to multicellular organization and organ structures [61], diseased organoids from patients' stem cells have been postulated to understand human pathological mechanism with the future aim of drug testing applications. For example, it includes cyctic fibrosis. Organoid swelling assay from patients' stem cells has been used to select the therapeutic compound in a specific pathological context. Thus, 3D disease organoids potentially accelerate the study of human disease pathology, opening up a possibility to model drug responses at the organoid level, rather than 2D cellular condition. Moreover, organoids can be ultimately indicated for transplantation to provide a functional tissue to replace the damaged functions in patients. However, there remain several challenges that need to be overcome before application in clinics. In this section, we will discuss the possible promise and challenges of current 3D cell culture models for therapeutic purpose.

9.3.2 3D Models for Drug Discovery

3D cell cultures including organoids will likely become a powerful strategy of the early drug discovery process. To bridge the gap between 2D culture and in vivo, a broad spectrum of 3D cell cultures has been applied to understand the mechanisms underlying different diseases. For example, 3D models have reconstructed heterogeneity in advancing tumour biology, as standard 2D models are inadequate to address questions regarding painless disease, metastatic colonization, dormancy, relapse, and the rapid evolution of drug resistance [61]. This section summarizes efforts to implement various 3D cell culture technologies for different drug discovery processes (Fig. 9.3).

9.3.2.1 Spheroid-Based Drug Screening

Aggregated spheroids have been tested primarily in experimental cancer research and to oncology drug screening. Multicellular spheroid cultures were firstly investigated in 1970 to recapitulate the functional phenotype of human tumour cells and their responses to radiotherapy [62]. Spheroids of cancer cell lines have been used to investigate different aspects of the cancer invasion process, including the invasion of cells in a 3D spheroid into the surrounding 3D ECM structure and interactions between endothelial and tumour cells [63]. For example, Xu et al. developed a single-cell quantitative assay to monitor the invasion of cells in a spheroid through 3D Matrigel. They found that EGF accelerates invasion in the colon cancer cell line HT-29, whereas vandetanib dose dependently inhibits invasion [64]. Phosphatase

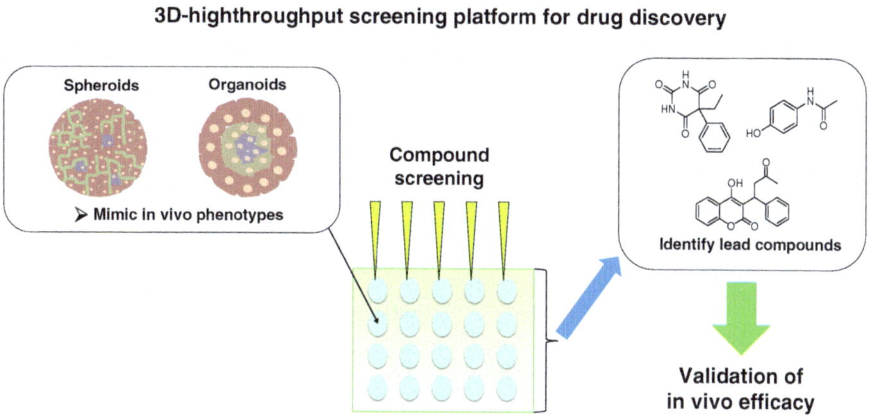

Fig. 9.3 Drug discovery using 3D cell culture systems. Spheroids or organoids are used in drug screens involving several thousand compounds, resulting in the identification of several compounds. Then, these compounds are evaluated for in vivo efficacy using appropriate animal models

and tensin homologue (PTEN) knockout increased the invasion rate of HCT116 cells in spheroids through 3D Matrigel, and the phosphatidylinositol-4,5-bisphosphate 3-kinase (PI3K) inhibitors LY294002 and wortmannin drastically reduced the invasiveness of the cells [65]. Senkowski et al. [66] screened 1600 compounds and identified five compounds that selectively targeted the hypoxic cell population using glucose-deprived multicellular tumour spheroids from colon cancer cell lines with inner hypoxia cultured in 384-well low-adhesion plates. Although 3D cell culture systems with HTS are still in their early stage and not all 3D cell culture models are appropriate for HTS or high-content screening (HCS), further attempts will be highly anticipated in developing simple protocols, high-density microplate formats, and compatibility with automation and multimode detection systems in a cost-effective manner.

9.3.2.2 Organotypic Culture- and Organoid-Based Drug Screening

Organotypic 3D coculture of cancer cells with other cell types has been used in high-throughput formats. Kenny et al. screened 2420 pharmacologically active compounds using a multilayered organotypic culture containing primary human fibroblasts, mesothelial cells, and ECM so as to reproduce the human ovarian cancer metastatic microenvironment [67]. Subsequent validation using secondary in vitro and in vivo assays confirmed two active compounds, β-escin and tomatine, which prevented ovarian cancer adhesion, invasion, and metastasis, leading to improved survival in mouse models. Thus, studies preliminarily showed the potential of complex 3D models to improve the relevance of drug screening assays.

Organoid cultures have also been applied to cancer models, as well as many other diseases including developmental disorders, infectious diseases, and neuronal degeneration [27]. Recently, Qian et al. developed brain region-specific organoids using mini-bioreactors for modelling Zika virus exposure [68]. A miniaturized spinning bioreactor makes the culture of organoids more cost effective, and this system may alleviate the issues of cost for organoid-based drug screening. However, full potential of organoid-based screen awaits the additional development of high-throughput and robust readout, reasonable fidelity, and compatibility to existing screening capability.

9.3.2.3 Organoid-Based Toxicology Study

The development of new drugs is enormously costly, mainly because of failures, particularly in late-stage clinical trials, which are in turn partially owing to unanticipated side effects, including cardiac and liver toxicity [69]. Therefore, there is considerable interest in developing approaches that could more effectively predict side effects in candidate drugs, thereby enabling the selection of candidates less likely to fail owing to toxicity in late-stage trials. For hepatotoxicity, human primary hepatocytes are widely used in order to complement the results of immortalized hepatic cell lines. However, these models also have limitations such as donor sources, rapid loss of function, and lot-to-lot variations. Recently, human ESC- and iPSC-derived hepatic cells were generated that express drug metabolism enzymes, such as cytochrome P450 3A4 (CYP3A4), that can take up indocyanine green and respond to known hepatotoxic drugs [70]. In addition, disease-specific human iPSC-derived hepatocytes offer a useful model for investigating the mechanism of inherited diseases and discovering specific drug treatments [71–73]. Beyond such drug discovery studies, iPSC-derived hepatocytes are expected to become a novel model for assessing drug safety. Serious adverse effects, such as liver failure, are major causes of drug attrition during clinical development [74] and withdrawal of marketed pharmaceuticals. In particular, drug-induced liver injury (DILI) is one of the major concerns in clinical practice. Although it is relatively uncommon, DILI is the leading cause of acute liver failure in the USA and a major reason for liver transplantation [75]. In preclinical studies, approximately 50% of candidate compounds present hepatic effects at supra-therapeutic doses in traditional laboratory animal-based models [76]. These models are not sufficiently predictive of DILI in humans [77]. Therefore, to overcome interspecies differences, primary human hepatocyte (PHH) spheroids have been considered potentially useful for modelling DILI pathologies [78], and human iPSC-derived hepatocyte-based models will also be useful for predicting DILI more accurately in humans compared to conventional models [22]; furthermore, these approaches could be adapted to high-throughput screening (HTS) systems in primary drug screens to select optimized drugs from large numbers of candidates [79, 80]. With the use of iPSC, human organoid technology would likely improve the accuracy and consistency by improving maturity, long-term persistence, and the microanatomical features such by forming bile canaliculi. Indeed, reports indicated reasonable similarities to in vivo organs and proven utility for studies of

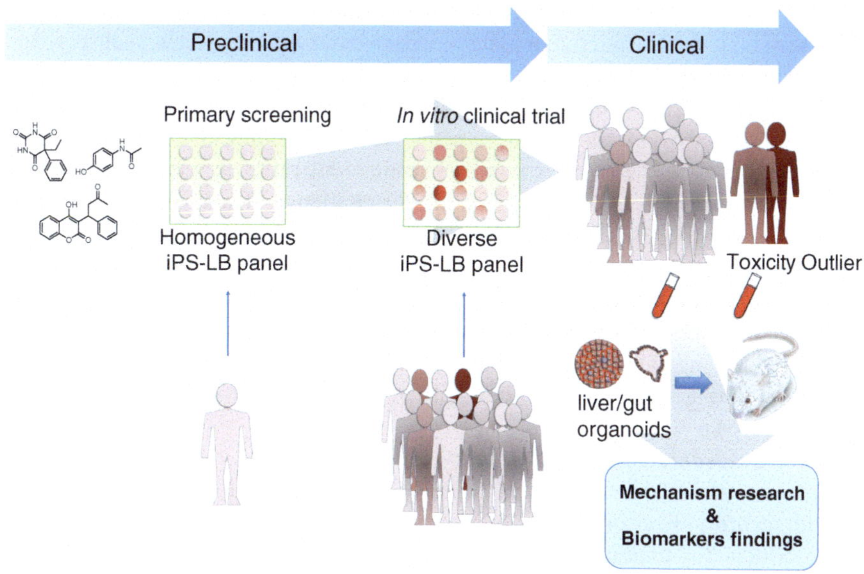

Fig. 9.4 A possible strategy to prevent side effects using organoids. In primary drug screening, iPSC-derived functional liver organoids are a useful model to predict side effects such as DILI using the HTS system. Specifically, iPSC organoids generated from diverse populations showing different drug susceptibilities would provide an effective model. Finally, animal models transplanted with iPSC liver organoids are expected to be useful to investigate the in-depth mechanism and provide biomarkers for risk stratification

nephrotoxicity [81] and cardiotoxicity testing [82]. Similar trends with liver organoids will allow DILI to be predicted more precisely through a deeper understanding of its mechanism and are expected to be applied as a cell panel for use in primary drug screening with the use of diverse population iPSC panels with different drug susceptibilities [22]. Thus, this model would be an effective tool for predicting idiosyncratic DILI observed in small populations and has been proposed to serve as a novel translational research model in preclinical phase (Fig. 9.4).

9.3.3 Challenges, Limitations, and Future Perspectives for 3D Cell Cultures

Many challenges still prevent the widespread application of 3D cell culture technologies for drug discovery. In fact, 3D screens are limited in their applications, because of the many difficulties currently associated with them, such as the following.

1. *High-throughput screening*. Many 3D cell models, such as organoids, have more complex morphology and function than 2D cultured cells, causing difficulties for their systematic assessments. This also presents challenges in standardization of

culture and assay protocols, phenotypes, and output data for analysis. The development of high-density microtiter plate-based spheroid-forming plates, such as 1536-well, low-adhesion spheroid plates, would be an attractive solution to streamline 3D spheroid-based drug screening. Recently, the 1536-well spheroid-forming plates also make high-throughput screening (HTS) economically viable [66].
2. *Improving phenotypic relevance*. Obtaining desirable in vivo drug effects from 3D screening approaches is a further challenge. Thus, it is necessary to bridge the gap between 2D culture and in vivo 3D culture system. Identifying a clinically relevant and measurable phenotype in 3D models is critical to rationalize and enhance the screening process. 3D cell culture predictive markers for drug efficacy and toxicity also need to be further determined and validated using existing data obtained in humans. Only a small set of data has confirmed that the efficacy and toxicity of drugs in 3D models resemble clinical data [61].
3. *Building advanced structures*. Improving innervation and vascularization of 3D culture models can recapitulate the complexity of human organ function [83]. With regard to liver organoids, bile canaliculi structures and transport functions are critical for toxicity study which is yet to be proven in organoids. Going forward, advancing organoid technologies is pivotal for recapitulating more complex microanatomy in vitro to achieve higher-order functions. Improving vascularization provides continuous nutrient supply from blood vessels and maturation for organoids. Coculturing with endothelial cells can generate vascular-like networks in organoids [26].

9.3.4 Emerging Approaches

Some of the most exciting recent advances in organoid research have come from synergistic interactions between biologists and bioengineers, resulting in a new strategy to control fluid flow, called organ-on-a-chip. Organ-on-a-chip can provide some solutions for the previously mentioned limitations of 3D cell cultures for disease modelling and drug screening. For instance, cultured human skin tissue in an organ-on-a-chip has been successfully used as a surrogate to model melanoma cancer growth [84]. When human melanoma cell lines were incorporated, the cultured skin tissue recapitulated the natural features of melanocyte homeostasis and melanoma progression in human skin. They displayed the same characteristics as the original tumour, vertical and radial growth phases, and metastatic melanoma cells, in vivo. The lung-on-chip was also developed to mimic breathing by stretching and compressing an artificial alveolar-capillary barrier using a cyclic vacuum machine. This was used to model pathogenic infections and inflammatory responses to air pollutants [85]. Similarly, an airway-on-a-chip device, lined with living human bronchiolar epithelium from normal or chronic obstructive pulmonary disease (COPD) patients, was connected to an instrument that can simulate exhalation of cigarette smoke and has chips to evaluate the effects of smoke-induced

Table 9.1 Comparison of the strengths and weakness of organoid and organ-on-a-chip methodologies

Features/technologies	Organoid	Organ-on-a-chip
Structural variability	High	Low
Cellular fidelity	High	Low
Flow input/output	Absent	Present
Built-in readouts	Absent	Present
Environmental control	Low	High

pathophysiology in vitro. Aref et al. developed an organ-on-a-chip consisting of lung cancer spheroids in a 3D matrix gel to recapitulate the epithelial-mesenchymal transition during cancer progression [86]. Their results indicate that, at least in the cell line model, organ-on-a-chip approach entails both qualitative and quantitative advantages in drug response over conventional 2D monolayer cells.

Organoid technology and organ-on-a-chip engineering have emerged as two distinct approaches for stem cell-derived 3D tissue preparation. To know the pros and cons of each method is vital for improving the use of 3D cell culture systems in disease modelling (Table 9.1). Considering structural variability and cellular fidelity, organ-on-a-chip devices reduce these sources of variation because their cellular composition is very simplified and contains only one or two cell types that have limited physiological relevancies. However, organ-on-a-chip approaches can control the geometry input and output flow conditions in addition to environmental conditions such as nutrient and oxygen supply as well as shear stress stimulation. Although organ-on-a-chip systems are amendable to high-throughput multiplexed sensing systems using microscopy, microfluorimetry, mechanical measurements, and multiple electrode arrays, they are subject to cellular fidelity challenges. Alternatively, organoid systems that achieve the higher-order function of 3D tissues and closely resemble in vivo organ tissues but generally lack screenable readouts. Recently, researchers analysed photosensitive cells in brain organoids by incorporating a high-density silicon microelectrode sensing system, illustrating the benefits of combining high-throughput readout sensors with organoids to further investigate complex cellular interactions [34]. Integrated approaches will thus facilitate the development of phenotypic screening platforms for drug discovery and development applications with automated functional readouts. Therefore, their strategic integration will potentially address the limitations of 3D culture models and provide a path towards a superior, synergistic strategy for constructing tissues that will truly deliver on the promise of regenerative and precision medicine [87].

9.4 Conclusion

In this review, the evolutionary history of 3D organoid models and their potential and challenges for drug discovery applications were discussed. Many tissue organoids have already been established using ES cell and iPS cells for drug screening and their toxicology test. Although tissue organoids have some native tissue function, they are still missing aspects of mature function partly due to oversimplified culture system lacking supportive structures such as vasculature. Additionally, improvement efficiency and reproducibility will be critical going forward. Multidisciplinary efforts including biologists, bioengineers, and professionals from the other emerging areas will greatly facilitate the therapeutic conversion process.

References

1. Wilson HV. A new method by which sponges may be artificially reared. Science. 1907;25:912–5.
2. Martz E, Phillips HM, Steinberg MS. Contact inhibitions of overlapping and differential cell adhesion: a sufficient model for the control of certain cell culture morphologies. J Cell Sci. 1974;16:401–19.
3. Takeichi M. Self-organization of animal tissues: cadherin-mediated processes. Dev Cell. 2011;21:24–6.
4. Li ML, Aggeler J, Farson DA, Hatier C, Hassell J, Bissell MJ. Influence of a reconstituted basement membrane and its components on casein gene expression and secretion in mouse mammary epithelial cells. Proc Natl Acad Sci U S A. 1987;84:136–40.
5. Shannon JM, Mason RJ, Jennings SD. Functional differentiation of alveolar type II epithelial cells in vitro: effects of cell shape, cell-matrix interactions and cell-cell interactions. Biochim Biophys Acta. 1987;931:143–56.
6. Montesano R, Schaller G, Orci L. Induction of epithelial tubular morphogenesis in vitro by fibroblast-derived soluble factors. Cell. 1991;66:697–711.
7. Gjorevski N, Sachs N, Manfrin A, Giger S, Bragina ME, Ordez-Morn P, Clevers H, Lutolf MP. Designer matrices for intestinal stem cell and organoid culture. Nature. 2016;539(7630):560–4.
8. Cruz-Acua R, Quirs M, Farkas AE, Dedhia PH, Huang S, Siuda D, Garca-Hernndez V, Miller AJ, Spence JR, Nusrat A, Garca AJ. Synthetic hydrogels for human intestinal organoid generation and colonic wound repair. Nat Cell Biol. 2017;19(11):1326–35.
9. Thomson JA, Itskovitz-Eldor J, Shapiro SS, Waknitz MA, Swiergiel JJ, Marshall VS, Jones JM. Embryonic stem cell lines derived from human blastocysts. Science. 1998;282:1145–7.
10. Takahashi K, Yamanaka S. Induction of pluripotent stem cells from mouse embryonic and adult fibroblast cultures by defined factors. Cell. 2006;126:663–76.
11. Eiraku M, Watanabe K, Matsuo-Takasaki M, Kawada M, Yonemura S, Matsumura M, Wataya T, Nishiyama A, Muguruma K, Sasai Y. Self-organized formation of polarized cortical tissues from ESCs and its active manipulation by extrinsic signals. Cell Stem Cell. 2008;3:519–32.
12. Sato T, Vries RG, Snippert HJ, van de Wetering M, Barker N, Stange DE, van Es JH, Abo A, Kujala P, Peters PJ, Clevers H. Single Lgr5 stem cells build crypt-villus structures in vitro without a mesenchymal niche. Nature. 2009;459:262–5.
13. Eiraku M, Takata N, Ishibashi H, Kawada M, Sakakura E, Okuda S, Sekiguchi K, Adachi T, Sasai Y. Self-organizing optic-cup morphogenesis in three-dimensional culture. Nature. 2011;472:51–6.
14. Spence JR, Mayhew CN, Rankin SA, Kuhar MF, Vallance JE, Tolle K, Hoskins EE, Kalinichenko VV, Wells SI, Zorn AM, Shroyer NF, Wells JM. Directed differentiation of human pluripotent stem cells into intestinal tissue in vitro. Nature. 2011;470:105–9.
15. Takebe T, Sekine K, Enomura M, Koike H, Kimura M, Ogaeri T, Zhang RR, Ueno Y, Zheng YW, Koike N, Aoyama S, Adachi Y, Taniguchi H. Vascularized and functional human liver from an iPSC-derived organ bud transplant. Nature. 2013;499:481–4.
16. Lancaster MA, Renner M, Martin CA, Wenzel D, Bicknell LS, Hurles ME, Homfray T, Penninger JM, Jackson AP, Knoblich JA. Cerebral organoids model human brain development and microcephaly. Nature. 2013;501:373–9.

17. Taguchi A, Kaku Y, Ohmori T, Sharmin S, Ogawa M, Sasaki H, Nishinakamura R. Redefining the in vivo origin of metanephric nephron progenitors enables generation of complex kidney structures from pluripotent stem cells. Cell Stem Cell. 2014;14:53–67.
18. Takasato M, Er PX, Becroft M, Vanslambrouck JM, Stanley EG, Elefanty AG, Little MH. Directing human embryonic stem cell differentiation towards a renal lineage generates a self-organizing kidney. Nat Cell Biol. 2014;16:118–26.
19. Si-Tayeb K, Noto FK, Sepac A, Sedlic F, Bosnjak ZJ, Lough JW, Duncan SA. Generation of human induced pluripotent stem cells by simple transient transfection of plasmid DNA encoding reprogramming factors. BMC Dev Biol. 2010;10:81.
20. Swistowski A, Peng J, Liu Q, Mali P, Rao MS, Cheng L, Zeng X. Efficient generation of functional dopaminergic neurons from human induced pluripotent stem cells under defined conditions. Stem Cells. 2010;28:1893–904.
21. Deng J, Shoemaker R, Xie B, Gore A, LeProust EM, Antosiewicz-Bourget J, Egli D, Maherali N, Park IH, Yu J, Daley GQ, Eggan K, Hochedlinger K, Thomson J, Wang W, Gao Y, Zhang K. Targeted bisulfite sequencing reveals changes in DNA methylation associated with nuclear reprogramming. Nat Biotechnol. 2009;27:353–60.
22. Shinozawa T, Yoshikawa HY, Takebe T. Reverse engineering liver buds through self-driven condensation and organization towards medical application. Dev Biol. 2016;420:221–9.
23. Huch M, Gehart H, van Boxtel R, Hamer K, Blokzijl F, Verstegen MM, Ellis E, van Wenum M, Fuchs SA, de Ligt J, van de Wetering M, Sasaki N, Boers SJ, Kemperman H, de Jonge J, Ijzermans JN, Nieuwenhuis EE, Hoekstra R, Strom S, Vries RR, van der Laan LJ, Cuppen E, Clevers H. Long-term culture of genome-stable bipotent stem cells from adult human liver. Cell. 2015;160(1–2):299–312.
24. Sampaziotis F, de Brito MC, Madrigal P, Bertero A, Saeb-Parsy K, Soares FAC, Schrumpf E, Melum E, Karlsen TH, Bradley JA, Gelson WT, Davies S, Baker A, Kaser A, Alexander GJ, Hannan NRF, Vallier L. Cholangiocytes derived from human induced pluripotent stem cells for disease modeling and drug validation. Nat Biotechnol. 2015;33(8):845–52.
25. Ogawa M, Ogawa S, Bear CE, Ahmadi S, Chin S, Li B, Grompe M, Keller G, Kamath BM, Ghanekar A. Directed differentiation of cholangiocytes from human pluripotent stem cells. Nat Biotechnol. 2015;33(8):853–61.
26. Takebe T, Enomura M, Yoshizawa E, Kimura M, Koike H, Ueno Y, Matsuzaki T, Yamazaki T, Toyohara T, Osafune K, Nakauchi H, Yoshikawa HY, Taniguchi H. Vascularized and complex organ buds from diverse tissues via mesenchymal cell-driven condensation. Cell Stem Cell. 2015;16:556–65.
27. Lancaster MA, Knoblich JA. Organogenesis in a dish: modeling development and disease using organoid technologies. Science. 2014;345:1247125.
28. Elkabetz Y, Panagiotakos G, Al Shamy G, Socci ND, Tabar V, Studer L. Human ES cell-derived neural rosettes reveal a functionally distinct early neural stem cell stage. Genes Dev. 2008;22:152–65.
29. Zhang SC, Wernig M, Duncan ID, Brüstle O, Thomson JA. In vitro differentiation of transplantable neural precursors from human embryonic stem cells. Nat Biotechnol. 2001;19:112–33.
30. Muguruma K, Nishiyama A, Kawakami H, Hashimoto K, Sasai Y. Self-organization of polarized cerebellar tissue in 3D culture of human pluripotent stem cells. Cell Rep. 2015;10(4):537–50.
31. Lancaster MA, Corsini NS, Wolfinger S, Gustafson EH, Phillips AW, Burkard TR, Otani T, Livesey FJ, Knoblich JA. Guided self-organization and cortical plate formation in human brain organoids. Nat Biotechnol. 2017;35(7):659–66.
32. Nakano T, Ando S, Takata N, Kawada M, Muguruma K, Sekiguchi K, Saito K, Yonemura S, Eiraku M, Sasai Y. Self-formation of optic cups and storable stratified neural retina from human ESCs. Cell Stem Cell. 2012;10:771–85.
33. Zhong X, Gutierrez C, Xue T, Hampton C, Vergara MN, Cao LH, Peters A, Park TS, Zambidis ET, Meyer JS, Gamm DM, Yau KW, Canto-Soler MV. Generation of three-dimensional retinal tissue with functional photoreceptors from human iPSCs. Nat Commun. 2014;5:4047.
34. Quadrato G, Nguyen T, Macosko EZ, Sherwood JL, Min Yang S, Berger DR, Maria N, Scholvin J, Goldman M, Kinney JP, Boyden ES, Lichtman JW, Williams ZM, McCarroll

SA, Arlotta P. Cell diversity and network dynamics in photosensitive human brain organoids. Nature. 2017;545(7652):48–53.
35. Weiss P, Taylor AC. Reconstitution of complete organs from single-cell supsensions of chick embryos in advanced stages of differentiation. Proc Natl Acad Sci U S A. 1960;46:1177–85.
36. Clevers H. Modeling development and disease with organoids. Cell. 2016;165:1586–97.
37. Mae SI, Shono A, Shiota F, Yasuno T, Kajiwara M, Gotoda-Nishimura N, Arai S, Sato-Otubo A, Toyoda T, Takahashi K, Nakayama N, Cowan CA, Aoi T, Ogawa S, McMahon AP, Yamanaka S, Osafune K. Monitoring and robust induction of nephrogenic intermediate mesoderm from human pluripotent stem cells. Nat Commun. 2013;4:1367.
38. Xia Y, Nivet E, Sancho-Martinez I, Gallegos T, Suzuki K, Okamura D, Wu MZ, Dubova I, Esteban CR, Montserrat N, Campistol JM, Izpisua Belmonte JC. Directed differentiation of human pluripotent cells to ureteric bud kidney progenitor-like cells. Nat Cell Biol. 2013;15:1507–15.
39. Samira M, Akiko M, Thomas CF, Sauveur SFJ, Mariko H, Tadanori M, Kristen R, Seyoon C, Richard N, Miles I, Tohid F, Sandeep K, James CW, George MC, Ingber DE. Mature induced-pluripotent-stem-cell-derived human podocytes reconstitute kidney glomerular-capillary-wall function on a chip. Nat Biomed Eng. 2017;1:0069.
40. Heath JK. Transcriptional networks and signaling pathways that govern vertebrate intestinal development. Curr Top Dev Biol. 2010;90:159–92.
41. Gjorevski N, Sachs N, Manfrin A, Giger S, Bragina ME, Ordóñez-Morán P, Clevers H, Lutolf MP. Designer matrices for intestinal stem cell and organoid culture. Nature. 2016;539(7630):560–4.
42. Miura S, Suzuki A. Generation of mouse and human organoid-forming intestinal progenitor cells by direct lineage reprogramming. Cell Stem Cell. 2017;21(4):456–71.
43. McCracken KW, Catá EM, Crawford CM, Sinagoga KL, Schumacher M, Rockich BE, Tsai YH, Mayhew CN, Spence JR, Zavros Y, Wells JM. Modelling human development and disease in pluripotent stem-cell-derived gastric organoids. Nature. 2014;516:400–4.
44. Bartfeld S, Bayram T, van de Wetering M, Huch M, Begthel H, Kujala P, Vries R, Peters PJ, Clevers H. In vitro expansion of human gastric epithelial stem cells and their responses to bacterial infection. Gastroenterology. 2015;148:126–136.e6.
45. McCracken KW, Aihara E, Martin B, Crawford CM, Broda T, Treguier J, Zhang X, Shannon JM, Montrose MH, Wells JM. Wnt/β-catenin promotes gastric fundus specification in mice and humans. Nature. 2017;541(7636):182–7.
46. Greggio C, de Franceschi F, Figueiredo-Larsen M, Gobaa S, Ranga A, Semb H, Lutolf M, Grapin-Botton A. Artificial three-dimensional niches deconstruct pancreas development in vitro. Development. 2013;140(21):4452–62.
47. Huch M, Bonfanti P, Boj SF, Sato T, Loomans CJ, van de Wetering M, Sojoodi M, Li VS, Schuijers J, Gracanin A, Ringnalda F, Begthel H, Hamer K, Mulder J, van Es JH, de Koning E, Vries RG, Heimberg H, Clevers H. Unlimited in vitro expansion of adult bi-potent pancreas progenitors through the Lgr5/R-spondin axis. EMBO J. 2013;32(20):2708–21.
48. Boj SF, Hwang CI, Baker LA, Chio II, Engle DD, Corbo V, Jager M, Ponz-Sarvise M, Tiriac H, Spector MS, Gracanin A, Oni T, Yu KH, van Boxtel R, Huch M, Rivera KD, Wilson JP, Feigin ME, Öhlund D, Handly-Santana A, Ardito-Abraham CM, Ludwig M, Elyada E, Alagesan B, Biffi G, Yordanov GN, Delcuze B, Creighton B, Wright K, Park Y, Morsink FH, Molenaar IQ, Borel Rinkes IH, Cuppen E, Hao Y, Jin Y, Nijman IJ, Iacobuzio-Donahue C, Leach SD, Pappin DJ, Hammell M, Klimstra DS, Basturk O, Hruban RH, Offerhaus GJ, Vries RG, Clevers H, Tuveson DA. Organoid models of human and mouse ductal pancreatic cancer. Cell. 2015;160(1–2):324–38.
49. Broutier L, Andersson-Rolf A, Hindley CJ, Boj SF, Clevers H, Koo BK, Huch M. Culture and establishment of self-renewing human and mouse adult liver and pancreas 3D organoids and their genetic manipulation. Nat Protoc. 2016;11(9):1724–43.
50. Mondrinos MJ, Jones PL, Finck CM, Lelkes PI. Engineering de novo assembly of fetal pulmonary organoids. Tissue Eng Part A. 2014;20:2892–907.

51. Rock JR, Onaitis MW, Rawlins EL, et al. Basal cells as stem cells of the mouse trachea and human airway epithelium. Proc Natl Acad Sci U S A. 2009;106:12771–5.
52. Dye BR, Dedhia PH, Miller AJ, Nagy MS, White ES, Shea LD, Spence JR. A bioengineered niche promotes in vivo engraftment and maturation of pluripotent stem cell derived human lung organoids. Elife. 2016;5:e19732.
53. Lee JH, Tammela T, Hofree M, Choi J, Marjanovic ND, Han S, Canner D, Wu K, Paschini M, Bhang DH, Jacks T, Regev A, Kim CF. Anatomically and functionally distinct lung mesenchymal populations marked by Lgr5 and Lgr6. Cell. 2017;170(6):1149–63.
54. Chen YW, Huang SX, de Carvalho ALRT, Ho SH, Islam MN, Volpi S, Notarangelo LD, Ciancanelli M, Casanova JL, Bhattacharya J, Liang AF, Palermo LM, Porotto M, Moscona A, Snoeck HW. A three-dimensional model of human lung development and disease from pluripotent stem cells. Nat Cell Biol. 2017;19(5):542–9.
55. McCauley KB, Hawkins F, Serra M, Thomas DC, Jacob A, Kotton DN. Efficient derivation of functional human airway epithelium from pluripotent stem cells via temporal regulation of Wnt signaling. Cell Stem Cell. 2017;20(6):844–57.
56. Hogan BLM, Barkauskas CE, Chapman HA, et al. Repair and regeneration of the respiratory system: complexity, plasticity, and mechanisms of lung stem cell function. Cell Stem Cell. 2014;15:123–38.
57. del Moral P-M, Warburton D. Explant culture of mouse embryonic whole lung, isolated epithelium, or mesenchyme under chemically defined conditions as a system to evaluate the molecular mechanism of branching morphogenesis and cellular differentiation. Methods Mol Biol. 2010;633:71–9.
58. Nikolić MZ, Rawlins EL. Lung organoids and their use to study cell-cell interaction. Curr Pathobiol Rep. 2017;5(2):223–31.
59. Zhang S, Zhou X, Chen T, et al. Single primary fetal lung cells generate alveolar structures in vitro. In Vitro Cell Dev Biol Anim. 2014;50:87–93.
60. Sucre JMS, Wilkinson D, Vijayaraj P, et al. A threedimensional human model of the fibroblast activation that accompanies bronchopulmonary dysplasia identifies Notch-mediated pathophysiology. Am J Physiol Lung Cell Mol Physiol. 2016;310:L889–98.
61. Fang Y, Eglen RM. Three-dimensional cell cultures in drug discovery and development. SLAS Discov. 2017;22:456–72.
62. Sutherland RM, McCredie JA, Inch WR. Growth of multicell spheroids in tissue culture as a model of nodular carcinomas. J Natl Cancer Inst. 1971;46:113–20.
63. Ghosh S, Joshi MB, Ivanov D, Feder-Mengus C, Spagnoli GC, Martin I, Erne P, Resink TJ. Use of multicellular tumor spheroids to dissect endothelial cell-tumor cell interactions: a role for T-cadherin in tumor angiogenesis. FEBS Lett. 2007;581:4523–8.
64. Febles NK, Ferrie AM, Fang Y. Label-free single cell kinetics of the invasion of spheroidal colon cancer cells through 3D Matrigel. Anal Chem. 2014;86:8842–9.
65. Chandrasekaran S, Deng H, Fang Y. PTEN deletion potentiates invasion of colorectal cancer spheroidal cells through 3D Matrigel. Integr Biol. 2015;7:324–34.
66. Senkowski W, Zhang X, Olofsson MH, Isacson R, Höglund U, Gustafsson M, Nygren P, Linder S, Larsson R, Fryknäs M. Three-dimensional cell culture-based screening identifies the anthelmintic drug nitazoxanide as a candidate for treatment of colorectal cancer. Mol Cancer Ther. 2015;14:1504–16.
67. Kenny HA, Lal-Nag M, White EA, Shen M, Chiang CY, Mitra AK, Zhang Y, Curtis M, Schryver EM, Bettis S, Jadhav A, Boxer MB, Li Z, Ferrer M, Lengyel E. Quantitative high throughput screening using a primary human three-dimensional organotypic culture predicts in vivo efficacy. Nat Commun. 2015;6:6220.
68. Qian X, Nguyen HN, Song MM, Hadiono C, Ogden SC, Hammack C, Yao B, Hamersky GR, Jacob F, Zhong C, Yoon KJ, Jeang W, Lin L, Li Y, Thakor J, Berg DA, Zhang C, Kang E, Chickering M, Nauen D, Ho CY, Wen Z, Christian KM, Shi PY, Maher BJ, Wu H, Jin P, Tang H, Song H, Ming GL. Brain-region-specific organoids using mini-bioreactors for modeling ZIKV exposure. Cell. 2016;165:1238–54.

69. Scannell JW, Blanckley A, Boldon H, Warrington B. Diagnosing the decline in pharmaceutical R&D efficiency. Nat Rev Drug Discov. 2012;11(3):191–200.
70. Shi Y, Inoue H, Wu JC, Yamanaka S. Induced pluripotent stem cell technology: a decade of progress. Nat Rev Drug Discov. 2017;16:115–30.
71. Choi SM, Kim Y, Shim JS, Park JT, Wang RH, Leach SD, Liu JO, Deng C, Ye Z, Jang YY. Efficient drug screening and gene correction for treating liver disease using patient-specific stem cells. Hepatology. 2013;57:2458–68.
72. Rashid T, Takebe T, Nakauchi H. Novel strategies for liver therapy using stem cells. Gut. 2015;64:1–4.
73. Tafaleng EN, Chakraborty S, Han B, Hale P, Wu W, Soto-Gutierrez A, Feghali-Bostwick CA, Wilson AA, Kotton DN, Nagaya M, Strom SC, Roy-Chowdhury J, Stolz DB, Perlmutter DH, Fox IJ. Induced pluripotent stem cells model personalized variations in liver disease resulting from α1-antitrypsin deficiency. Hepatology. 2015;62:147–57.
74. Hay M, Thomas DW, Craighead JL, Economides C, Rosenthal J. Clinical development success rates for investigational drugs. Nat Biotechnol. 2014;32:40–51.
75. Reuben A, Koch DG, Lee WM, Acute Liver Failure Study Group. Drug-induced acute liver failure: results of a U.S. multicenter, prospective study. Hepatology. 2010;52:2065–76.
76. Amacher DE. Serum transaminase elevations as indicators of hepatic injury following the administration of drugs. Regul Toxicol Pharmacol. 1998;27:119–30.
77. Navarro VJ, Senior JR. Drug-related hepatotoxicity. N Engl J Med. 2006;354:731–9.
78. Bell CC, Hendriks DF, Moro SM, Ellis E, Walsh J, Renblom A, Fredriksson Puigvert L, Dankers AC, Jacobs F, Snoeys J, Sison-Young RL, Jenkins RE, Nordling Å, Mkrtchian S, Park BK, Kitteringham NR, Goldring CE, Lauschke VM, Ingelman-Sundberg M. Characterization of primary human hepatocyte spheroids as a model system for drug-induced liver injury, liver function and disease. Sci Rep. 2016;6:25187.
79. Sjogren AK, Liljevald M, Glinghammar B, Sagemark J, Li XQ, Jonebring A, Cotgreave I, Brolén G, Andersson TB. Critical differences in toxicity mechanisms in induced pluripotent stem cell-derived hepatocytes, hepatic cell lines and primary hepatocytes. Arch Toxicol. 2014;88:1427–37.
80. Krueger W, Boelsterli UA, Rasmussen TP. Stem cell strategies to evaluate idiosyncratic drug-induced liver injury. J Clin Transl Hepatol. 2014;2:143–52.
81. Morizane R, Lam AQ, Freedman BS, Kishi S, Valerius MT, Bonventre JV. Nephron organoids derived from human pluripotent stem cells model kidney development and injury. Nat Biotechnol. 2015;33:1193–200.
82. Ranga A, Gjorevski N, Lutolf MP. Drug discovery through stem cell-based organoid models. Adv Drug Deliv Rev. 2014;69–70:19–28.
83. Workman MJ, Mahe MM, Trisno S, Poling HM, Watson CL, Sundaram N, Chang CF, Schiesser J, Aubert P, Stanley EG, Elefanty AG, Miyaoka Y, Mandegar MA, Conklin BR, Neunlist M, Brugmann SA, Helmrath MA, Wells JM. Engineered human pluripotent-stem-cell-derived intestinal tissues with a functional enteric nervous system. Nat Med. 2017;23(1):49–59.
84. Li L, Fukunaga-Kalabis M, Herlyn M. The three-dimensional human skin reconstruct model: a tool to study normal skin and melanoma progression. J Vis Exp. 2011;54:2937.
85. Huh D, Matthews BD, Mammoto A, Montoya-Zavala M, Hsin HY, Ingber DE. Reconstituting organ-level lung functions on a chip. Science. 2010;328:1662–8.
86. Aref AR, Huang RY, Yu W, Chua KN, Sun W, Tu TY, Bai J, Sim WJ, Zervantonakis IK, Thiery JP, Kamm RD. Screening therapeutic EMT blocking agents in a three-dimensional microenvironment. Integr Biol. 2013;5:381–9.
87. Takebe T, Zhang B, Radisic M. Synergistic engineering: organoids meet organs-on-a-chip. Cell Stem Cell. 2017;21:297–300.

Chapter 10
In Vivo Cell Conversion as a New Cell Therapy

Hedong Li, Lei Zhang, Yuchen Chen, Zheng Wu, Zhuofan Lei, and Gong Chen

Abstract Cells are fundamental functioning units of the human body. Every day there are hundreds and thousands of cells die or are replaced in human beings. Some organs such as the skin and liver have highly effective self-repair mechanisms to replenish the lost cells with local progenitor cells, but other organs such as the brain and spinal cord do not have much self-repair capability, particularly in the adult stage. Classically, cell therapy is meant to supply new cells to the organ by transplanting external cells, typically stem cells that can be expanded in in vitro cultures. The induced pluripotent stem cells (iPSCs) provide an unprecedented cell source that can be derived directly from patients for tissue repair. In this review, we summarize a new technology called "*in vivo* cell conversion" that makes use of endogenous cells to trans-differentiate into another type of cells to replenish the lost cells caused by injury or diseases. The advantage of in vivo cell conversion is to relieve the burden of in vitro cell culture that is often difficult to control in terms of batch-to-batch variation and the risk of immunorejection after transplantation. In vivo cell conversion has been successfully achieved in the brain, spinal cord, retina, heart, liver, and pancreas. With the advent of sequencing technology that can decode the transcription factors critical for cell fate determination, it is in theory possible to convert any cells into their neighboring cells in vivo, with a right combination of transcription factors. Because it is using patients' own internal cells as the resource for tissue repair, we predict that in vivo cell conversion will be the next-generation cell therapy for regenerative medicine.

Keywords In vivo cell conversion · Neural repair · Neuroregeneration
Cell therapy · Gene therapy

H. Li (✉) · L. Zhang · Y. Chen · Z. Wu · Z. Lei · G. Chen (✉)
Department of Biology, Huck Institutes of Life Sciences, The Pennsylvania State University, University Park, PA 16802, USA
e-mail: hul40@psu.edu; gongchen@psu.edu; http://bio.psu.edu/directory/guc2

H. Inoue, Y. Nakamura (eds.), *Medical Applications of iPS Cells*,
Current Human Cell Research and Applications,
https://doi.org/10.1007/978-981-13-3672-0_10

10.1 Classical Cell Therapy for Neural Repair

Regenerative medicine requires creative approaches for translational purpose since many human organs have limited self-regeneration capability under physiological or pathological conditions. One such approach currently adopted for regenerative medicine is stem cell therapy. The promise of stem cells is to supply target cells upon transplantation in order to rescue the deficits induced by injury or diseases. One successful example is the hematopoietic stem cell, which has been transplanted into patients by way of bone marrow, peripheral blood, or umbilical cord blood to treat hematological disorders and malignancies such as leukemia. At least in mice, it has been demonstrated that a single hematopoietic stem cell can give rise to multi-lineages of blood cell types and reconstitute the lymphohematopoietic system [1]. Currently, there are many clinical trials using stem cells as therapies, but they also meet with challenges at the same time [2].

In the central nervous system (CNS), while neural stem cells (NSCs) can be efficiently induced to differentiate into different subtypes of neurons in cell cultures, they have largely failed to differentiate in vivo after transplantation due to non-neurogenic environment in injured or diseased adult CNS. In fact, most transplanted neural stem cells differentiate into glial cells (mainly astrocytes) in adult or injured CNS [3]. In several studies, the transplanted neural progenitor cells (NPCs) are suggested to benefit functional recovery after nerve injury through secreting neurotrophic factors that promote endogenous axonal regeneration rather than replacing lost neurons [4–6].

One successful case in regard to neuronal replacement is the transplantation of human fetal tissue into the striatum of Parkinson patients [7]. Another example in animal models is the transplantation of GABAergic interneuron progenitors [8, 9]. Previous reports have demonstrated that transplantation of MGE-derived GABAergic interneuron progenitors functionally improve neurological behavior in several animal disease models [10–14]. The transplantation of GABAergic interneurons into the spinal cord has also been demonstrated to reduce neuropathic pain that is induced by the injury [15]. These promising results enlighten hope to develop these treatments into clinical therapies, but they also face the common challenges for all cell replacement approaches—the source of transplantable human cells for therapies. Human embryonic stem cells (ESCs) [16] have been successfully differentiated into neuronal subtypes in vitro [17–20]. However, when transplanted into in vivo rodent models, human ESCs cannot be efficiently differentiated into target neurons in the CNS. Furthermore, human ESCs suffer from immunorejection after transplantation. Thus, most hESC studies were performed either in immunodeficient mice or with the use of immune suppressors. The autologous induced pluripotent stem cells (iPSCs) might be able to bypass immunorejection issues since they can be established from patient's own cells (see details in the next section). Finally, both hESCs and hiPSCs will have to deal with safety issues when come down to therapy, since both types of stem cells have the potential to form tumors in vivo.

An alternative strategy for the CNS injury repair is to make use of endogenous neural stem cells to replenish the lost neurons caused by the injury. The subventricular zone (SVZ) of the adult forebrain [21, 22] and subgranular zone (SGZ) of the hippocampus [23] are the two discrete brain regions where neural stem cells reside in mice [24] and human [25, 26]. Under physiological condition, neural stem cells in the SVZ is responsible for sending new neurons to the olfactory bulb via the rostral migratory stream (RMS) for new sense of smell [27–31], while SGZ neural stem cells give rise to new granule neurons in the hippocampus where they contribute to learning and memory [32–34]. This ongoing adult neurogenesis is a dynamically regulated process, since neuroregeneration is enhanced when animals are exposed to an enriched environment [35], novel stimuli [36], or subjected to voluntary exercise [37].

Upon injury or under disease conditions, NSC proliferation and neurogenesis can be promoted in the "niches" of the adult brain [38]. The newly generated neurons or neuroblasts can be redirected from their normal routes toward the site of injury [39, 40]. Previous reports have demonstrated that endogenously generated neurons can mature and integrate into the local circuits [41–43]. Besides well-documented NSCs in the SVZ and SGZ, scattered reports also find some newborn neurons in other brain regions [44], but in vivo lineage tracing studies mainly find glial cells derived from endogenous NSCs [45]. Most recent studies only found a few thousands neuroprogenitors in adult human brains, which have billions of neurons [46, 47]. Therefore, it is generally believed that endogenous neurogenesis may have limited capacity for injury repair, because many newborn neurons undergo apoptosis after generation [39, 48].

The induced pluripotent stem cell (iPSC) technology has revolutionized regenerative medicine. Reprogramming technology can be traced back to 1962, when Sir John Gurdon created a cloned frog with "nuclear transfer" technique [49], where the nucleus of a terminally differentiated somatic cell can be "reprogrammed" back to the primitive state of development equivalent to a fertilized egg. After three decades, the first mammal Dolly sheep was successfully cloned [50]. Most recently, cloned monkeys were also reported [51]. In cell cultures, fibroblast cells were reprogrammed into myoblasts with a single transcription factor MyoD1 [52]. However, this concept of cellular reprogramming has not been fully appreciated until Shinya Yamanaka and colleagues successfully reprogrammed mouse and human skin fibroblast cells into induced pluripotent stem cells using four transcription factors (Oct4, Sox2, Klf4, and c-Myc) [53, 54]. This iPSC technology makes it possible to reprogram patients' own cells into stem cells, and then differentiate into a variety of somatic cells for cell replacement therapies. Following this groundbreaking research, many laboratories around the world have generated iPSCs using different transcription factors [55–59], noncoding RNAs [60–63], and small molecules [64–66]. In addition to fibroblasts, iPSCs have been derived from a variety of cell types including liver and stomach cells [67], pancreatic cells [68], B lymphocytes [56], neural progenitors [69], T cells [70], and cord blood cells [71]. The iPSCs resemble, to a large extend, the naïve ESCs and can be expanded in cell cultures before differentiated into different cell types in a controlled manner. Importantly, iPS cells can be derived from patient's own cells to overcome immunorejection when used for transplantation [72–77]. However, iPS cells still suffer potential problems such as tumorigenesis after transplantation [78]. Careful evaluation in the clinical setting is much needed before large-scale clinical trials [79, 80].

10.2 In Vivo Cell Conversion as the Next-Generation Cell Therapy

To overcome the limitation of cell transplantation using in vitro cultured cells, recent progress on in vivo reprogramming technology provides a new avenue for regenerative medicine using endogenous cells for tissue repair. Dr. Douglas Melton and colleagues first reported direct reprogramming of non-insulin-secreting cells into insulin-secreting β cells [81]. This was quickly followed with successful in vivo reprogramming in the heart [82–84], brain and spinal cord [85–93], and liver [94, 95]. The advantage of in vivo reprogramming is to make use of internal cells to regenerate target cells without any complication of transplanting external cells. However, so far the efficiency of in vivo reprogramming achieved in different organs is quite different, with the highest efficiency of 90% reported in adult mouse cortex [88]. The following sections review the success of in vivo reprogramming in each organ, with in-depth review in the CNS.

10.3 Pancreas

There are approximately three million cell clusters called pancreatic islets in the human pancreas [96]. Within these islets are two major types of cells involved in the regulation of blood glucose levels. The α cells secrete glucagon to increase glucose in the blood, while the β cells secrete insulin to decrease it. Chronic autoimmune-induced β-cell degeneration leads to a shortage of insulin and develops into type 1 diabetes (T1D) [97]. To correct the disease phenotype, pancreatic transplantation and boosting β-cell proliferation provide potential strategies to treat T1D [98]. Recently, in vivo reprogramming technology provides a novel strategy to generate functional β cells inside the pancreas from other internal cell types to treat T1D. Insulin-secreting β cells are located within the islets alongside other hormone-secreting cells including glucagon-secreting α cells and somatostatin-secreting δ cells. The α cells are a particularly attractive internal cell source for reprogramming, because they are developmentally related to β cells. Reprogramming α cells to β cells not only reduces glucagon but also increases insulin, both of which will work to reduce the glucose concentration in the blood. To achieve α to β cell conversion, master transcription factors involved in normal pancreatic cell development are investigated. Overexpressing a set of three transcription factors (Ngn3, Pdx1, and

Mafa) in the abundant pancreatic exocrine cells can convert exocrine cells into β cells in situ in the pancreases of adult diabetic mice [81]. Importantly, the newly induced β cells can secrete active insulin and ameliorate hyperglycemia in the chemically induced pancreatic injury mouse model [81]. In addition, Pax4 promotes the β and δ cell lineages, whereas Arx is involved in specifying cell fate of α cells. Either overexpression of Pax4 [99] or knockdown of Arx [100] promotes α cells to β-like cell conversion. Besides genetic methods, fasting-mimicking diets in mice induce Ngn3-driven regeneration of insulin-secreting β cells, which in turn reverses diabetic symptoms in mouse models [101]. Interestingly, enhancement of $GABA_A$ receptor signaling via long-term administration of GABA [102] or artemisinin [103] results in the conversion of glucagon-secreting α cells into functional insulin-secreting β-like cells in vivo. Because artemether is a widely deployed antimalarial drug and GABA is an additive in food or beverages, it will be interesting to conduct clinical trials to test their effects in T1D patients.

10.4 Heart

Cardiovascular diseases stand as the leading cause of human death globally. Besides whole heart transplantation, which depends heavily on the limited availability of donated organs, cardiac regenerative medicine holds promise to treat heart diseases. Different strategies based on various animal models were developed to replace the lost cardiomyocytes (CMs) with induced cardiomyocytes (iCMs). The CMs for transplantation may be derived from cardiac progenitor cells [104], bone marrow cells [105], and pluripotent stem cells (PSCs) [106]. iPS cells have been widely used in differentiation into iCMs, with some favorable outcomes in both structural and functional repair after iCM engraftment [107–110].

Recently, direct in vivo reprogramming technology opens a new avenue for regeneration in the heart [82–84]. The adult mammalian heart consists of cardiomyocytes and other cell types, including vascular cells and cardiac fibroblasts (CFs). The CFs comprise nearly half of the whole cardiac tissue and can proliferate and secrete growth factors upon cardiac injury. Thus, direct reprogramming of endogenous CFs into functional iCMs in situ seems to be an ideal approach for heart regeneration. Based on earlier studies [111–114], Ieda et al. screened 14 candidate genes that could activate cardiac programming in the CFs and identified 3 cardiac transcription factors (Gata4, Mef2c, and Tbx5, briefly as GMT) to directly transdifferentiate neonatal CFs into cardiomyocyte-like cells in vitro and in vivo [82].

In 2012, several groups independently made important progress on cardiac reprogramming in mouse myocardial infarction (MI) models in vivo [83, 84, 115, 116]. Olson and colleagues used GMT plus Hand2 (GHMT) to reprogram adult mouse CFs into beating cardiac-like myocytes in vitro and convert non-CMs into functional iCMs in vivo, showing calcium transients and action potentials [84]. Srivastava and colleagues achieved similar iCM conversion after GMT delivery by retrovirus into the peri-infarct areas of the mouse myocardium in vivo and observed beneficial effects after the treatment [83]. Furthermore, through lineage tracing strategy, they indicated that the converted cells originated mainly from CFs infected by GMT with ~12% conversion efficiency [83]. Another study used a combination of four microRNAs by lentiviral delivery into the ischemic mouse myocardium converted CFs into CM-like cells in vivo [116]. Later, they proved that the converted iCMs in vivo by microRNAs were comparable to those by transcription factors in morphological, physiological, and functional aspects [117].

Inspired by these studies, additional transcription factors [118, 119], microRNAs [117, 120], and small chemical molecules [119, 121] were explored to achieve CF-to-CM conversion with better reprogramming efficiency (from 3- to 50-folds in different instances) and functional maturity. In 2014, Ding and colleagues combined a single transcription factor (Oct4) with a chemical cocktail of SB431542 (TGFβ inhibitor), CHIR99021 (Wnt activator), Parnate and Forskolin to reprogram fibroblasts into iCMs successfully with ~1% conversion efficiency in vitro [119]. More recently, Srivastava and colleagues demonstrated that cardiac reprogramming in vivo by GMT could be greatly improved by intraperitoneal injection of SB431542 and XAV939 (Wnt inhibitor), which converted CFs into iCMs with fourfold higher ratio and more molecular similarity to primitive CMs and better cardiac function recovery after MI [122]. In a 2018 study by Ieda and colleagues, a non-integrating *Sendai virus* was used to deliver reprogramming factors, which remarkably increase the efficiency of CF-to-CM conversion by robust gene expression over retrovirus both in cultured mouse and human fibroblasts as well as in mouse model in vivo [123]. These studies suggest that specific genetic and epigenetic barriers may be the key obstacles toward successful cardiac reprogramming [121, 124, 125]. Compared with the use of viral vectors which may cause immune response [126], small molecules may facilitate the therapeutic application of in vivo cardiac reprogramming [122, 127].

On the other hand, CM renewal is not the only component required for heart regeneration. Supporting blood circulation is also necessary for metabolic recovery in infarcted myocardium. Vascularization via vascular endothelial growth factors (VEGF) in the scar tissue prior to GMT treatment could reduce myocardial fibrosis and improve CM conversion and myocardial function both in vitro and in vivo [128, 129]. Therefore, a favorable microenvironment can have a positive impact on cardiac reprogramming in situ.

Besides the encouraging results in murine models in vivo, cardiac reprogramming in human cells in vitro also made significant progress. Human fibroblasts (HFs) have been reprogrammed into iCMs in vitro with different combinations of transcription factors, microRNAs, and small molecules [130–132]. For example, a combination of five transcription factors (GMT plus ESRRG and MESP1) was used to directly convert HFs into iCMs [133]. Then, a combination of modified GHMT and three cytokines (BMP4, activin A, and bFGF) was capable of reprogramming HFs into induced cardiac progenitor cells, which could further differentiate into three cardiac lineages in vitro and improve cardiac function after transplantation in vivo [134]. In a more recent research, a cocktail of nine chemicals (9C) was reported to convert HFs into functional iCMs in vitro [135]. After transplantation into mouse infarcted hearts, the 9C-treated HFs were compatible with the injured heart environment and could further mature into CMs in vivo [135]. These findings raise the hope of in vivo reprogramming non-CMs into CMs for human heart regeneration, especially using pharmacological agents.

10.5 Liver

Liver is the only visceral organ that has a remarkable capacity to regenerate after either chemical injury or surgical removal [136]. However, its self-regeneration is still not enough to revert chronic hepatic diseases such as liver cirrhosis [137]. Under chronic liver injury, the myofibroblast will be over-activated and start to secrete excessive collagen into the tissue matrix, which significantly reduces hepatic blood flow and compromises hepatocyte function, ultimately leading to cirrhotic liver disease and liver failure [137]. So far the most effective treatment for the late stage liver fibrosis is liver transplantation. Recently, in vivo reprogramming technology provides an alternative approach to treat liver cirrhosis. Ectopic overexpression of four human transcription factors (FOXA3, HNF1A, HNF4A, and GATA4) reprogrammed liver myofibroblasts into hepatocytes in vitro [95], while overexpression of six murine transcription factors (hnf1a, hnf4a, foxa1, foxa2, foxa3, and foxa4) also converted myofibroblasts into hepatocytes in vitro [94]. After injection of the polycistronic viruses expressing multiple transcription factors into the mouse liver in vivo, the myofibroblasts were directly reprogrammed into functional hepatocytes, but both strategies suffered from very low reprogramming efficacy (<1% of total hepatocytes) [137]. Despite the functional hepatocytes and proper hepatic gene expression profile, the converted cells continued to express some of the myofibroblast genes, suggesting that they are an intermediate cell type that still retains some properties of myofibroblasts. Since activated myofibroblasts secrete large amount of collagen to destroy the normal functions of the liver and lead to cirrhosis, in vivo conversion of myofibroblasts not only generates new hepatocytes but also reduces harmful myofibroblasts. Therefore, although only a small number of myofibroblasts were converted into hepatocyte-like cells, both of the aforementioned conversion methods showed significant therapeutic benefits by reducing the extent of liver injury and fibrosis in several liver disease mouse models [94, 95].

10.6 In Vivo Cell Conversion in the CNS

In vivo reprogramming inside the CNS, including both the brain and spinal cord, has advanced rapidly over the past several years, largely due to the high efficiency of astrocyte-to-neuron conversion achieved through expressing a single or a combination of transcription factors. In some cases, the in vivo reprogramming efficiency can reach as high as 90% [88], making it exceptional in the entire regenerative medicine field. Such high in vivo reprogramming efficiency raises new hope of brain repair using in vivo neuroregeneration to treat neurodegenerative disorders without transplantation of external stem cells.

10.7 Cerebral Cortex

Neuroregeneration in the cerebral cortex has been proven very difficult in the past because of a lack of adult neural stem cells. Studies have shown very few newborn neurons in the cortex even in pathological conditions such as stroke [39] but rather reactive astrogliosis after cortical injury [138]. Molecular manipulations such as antagonizing Olig2 function or overexpressing neurogenic factor Pax6 can generate some immature neurons in injured cortex [139]. In cultured cortical astrocytes, overexpression of neurogenin-2 (Ngn2) can efficiently convert astrocytes into neurons in vitro with 90% efficiency [140]. However, when retroviruses expressing Ngn2 were injected into the adult mouse cortex, the in vivo reprogramming efficiency was very low, suggesting that in vitro and in vivo reprogramming can be very different. In contrast, when retroviruses expressing NeuroD1 were injected into the adult mouse cortex, many reactive astrocytes were directly converted into functional neurons, which had been verified with electrophysiological recordings [88]. Moreover, expression of NeuroD1 in reactive astrocytes in an adult (14-month-old) mouse model for Alzheimer's disease also directly converted reactive astrocytes into functional neurons [88]. Both Ngn2 and NeuroD1 belong to the same bHLH family of transcriptional factors that play an important role during early brain development. Why do they show such difference in in vivo reprogramming efficiency in the adult mouse cortex? Different from Ngn2, which is mainly functioning in embryonic stages, NeuroD1 also plays a role in the adult neurogenesis in hippocampus [141] and olfactory bulb [142]. This may partially explain why NeuroD1 acts more efficiently than Ngn2 in the adult brain for in vivo reprogramming. While Ngn2 alone is not very effective, it can convert astrocytes into neurons in the mouse cortex when combined with additional factors such as Bcl-2 [86, 143] or a high dose of fibroblast growth factor 2 (FGF2) and epidermal growth factor (EGF) [87]. However, it is unclear whether these Ngn2-converted neurons are fully functional due to a lack of electrophysiological analysis. In addition, another bHLH factor Ascl1 has been reported to successfully convert astrocytes into functional neurons in the midbrain, verified with electrophysiological recordings [89].

Besides astrocytes, NG2 glia in the cortex is also a potential candidate for in vivo conversion because of their remarkable self-renewal capacity [87, 88, 144]. Ectopic expression of NeuroD1 in NG2 glia has shown successful NG2-to-neuron conversion in the adult mouse cortex [88]. Interestingly, while the majority of astrocyte-converted neurons are glutamatergic, a small portion of NG2 glia-converted neurons are GABAergic [88], suggesting that different lineages of glial cells can be used to generate different subtypes of neurons for the treatment of various neurological disorders.

It is worth mentioning that new technology using CRISPR-dCas9 gene editing has generated transgenic mice that can target multiple transcription factors in the bHLH family to convert astrocytes into neurons, although the efficiency is relatively low [145].

10.8 Striatum

Striatum receives glutamatergic and dopaminergic input from the cortex and midbrain and modulates diverse brain functions. Disrupting neuronal circuits in the striatum may cause severe malfunction in motor, rewarding, cognitive, and emotional systems, as observed in Parkinson's disease (PD), Huntington's disease (HD), and ischemic stroke. GABAergic medium spiny neurons (MSNs) are the largest neuronal population in striatum sending their projections to multiple brain regions, and their activity is modulated by dopaminergic input as well as local interneurons. Although physically located next to neurogenic subventricular zone (SVZ), striatum is generally considered as a quiescent brain region with very limited neurogenesis through adulthood. However, newborn neurons have been observed in the striatum following ischemic stroke [39, 146–151]. While most of these newborn neurons are generated through neuroblasts that migrate from the SVZ toward injured areas, a small portion might be trans-differentiated from striatal astrocytes, a process regulated by the Notch signaling pathway [152]. However, the poor functional recovery after stroke suggests that such internal brain plasticity is not enough for self-repair. To overcome this limitation, Dr. Zhang and colleagues first reported the use of Sox2, a transcription factor highly expressed in neural stem cells, to convert striatal astrocytes into neuroblasts [91]. Those Sox2-converted neuroblasts behave similar to the endogenous ones showing high expression level of Ascl1 and DCX, and self-renewing capability [90]. When treated with BDNF and noggin or histone deacetylase inhibitor VPA, Sox2-converted neuroblasts are able to differentiate into functional neurons and integrate into neural circuits [91]. Interestingly, microRNA miR-302/367 combined with VPA also induces the conversion of striatal glial cells into DCX+ neuroblasts [153]. An advantage of generating neuroblasts is that they retain the proliferative capability, which may continuously generate new neurons to a significant number for repair. On the other hand, if the proliferation of neuroblasts cannot be precisely controlled, they might result in tumor formation. To circumvent this issue, several groups have employed bHLH transcription factors in combination with other factors to directly convert striatal astrocytes into neurons, including Ascl1 [89], Ascl1 plus Brn2 and Mytl1 [154], and Ngn2 plus growth factors FGF2 and EGF [87]. When Sox2 and Ascl1 were co-expressed in cortical NG2 glia, immature neuroblasts were generated, and a few of them displayed functional property of mature neurons as revealed by electrophysiological recordings [155].

Two recent studies reported that microRNAs [156] and nanoparticles [157] might facilitate in vivo neuronal reprogramming of dopaminergic (DA) neurons from striatal astrocytes. Specifically, miR-218, a microRNA regulating DA neuron development, enables transcription factors (ASCL1, LMX1A, and NEUROD1) to generate DA neuron from astrocytes in the striatum of a PD mouse model [156]. Although the converted neurons were relatively small in number (14.6 ± 8.5 DA neurons/brain section), they appeared to improve some performance in behavioral tests such as rotation, gait, and coordination. In the presence of defined electromagnetic fields, electromagnetized gold nanoparticles were reported to facilitate the transcription factors (Ascl1, Pitx2, Lmx1a, and Nurr1) in converting mouse midbrain astrocytes into DA neurons in a PD mouse model through activating the histone H3K27 acetylation [157]. Systemic delivery of AAV particles has been tested for in vivo reprogramming as well, as demonstrated by intravascular injection of AAV expressing NEUROD1 that can cross the blood-brain barrier (BBB) and convert a small number of astrocytes into neurons (around ten converted neurons per striatum) in postnatal mouse striatum [85]. Lineage tracing results suggest that these neurons are likely converted from nonreactive astrocytes, which may explain why the conversion efficiency is much lower than that converted from reactive astrocytes after injury [88].

Astrocytes are not a homogenous population and rather a diverse population with a variety of lineages in different brain and spinal cord regions [143, 158]. The heterogeneity of glial cells and distinct microenvironment suggest that glia-to-neuron conversion will have regional difference even using the same conversion factors [87]. Besides astrocytes, NG2 glia may serve as another cell source for in vivo neuronal reprogramming. A number of transcription factors have been tested in attempt to convert NG2 glia into neurons. The combinations of transcription factors, i.e., ALN (Ascl1, Lmx1a, Nurr1), Ng2ND1 (Ngn2, NeuroD1), NgAN (Ngn2, Ascl1, Nurr1), and NgLN (Ngn2, Lmx1a, Nurr1), all successfully converted striatal NG2 glia into neurons [159, 160]. More research is undergoing to generate specific subtypes of neurons in specific brain regions.

10.9 Spinal Cord

In vivo reprogramming has not been extensively studied in the spinal cord comparing to the brain. Spinal cord has very limited neurogenic capacity under both physiological and injured conditions [161–163]. While neural progenitors isolated from injured spinal cord are able to differentiate into neurons and glia in cell culture condition [164, 165], such neuronal differentiation rarely occurs in the injured spinal cord in vivo. Furthermore, transplanted multipotent neural stem cells end up with mostly glial cells in the injured spinal cord, supporting the idea that injury environment favors gliogenesis rather than neurogenesis in the spinal cord [166, 167]. To overcome this limitation, manipulation of microenvironment and cellular intrinsic properties has been explored to facilitate neuronal differentiation. In transected spinal cord, injection of a combination of growth factors (FGF2 and EGF) substantially increased the proliferation of neural progenitors (Olig2+, Nkx2.2+, and NG2+) and neuronal differentiation [163]. However, the newborn neurons appeared to stall at immature stage and only started to express mature neuronal markers such as NeuN, MAP 2, and synaptic marker synaptophysin after combined with overexpressing the neurogenic transcription factor NGN2. These newly added neurons, albeit substantial in number, have poor long-term survival even in the presence of BDNF, a neurotrophic factor that appears to help improve short-term survival of these cells [163].

More recently, Zhang lab reported using transcription factor Sox2 to reprogram astrocytes into neuroblasts before further differentiating into neurons in the spinal cord [92, 93]. Sox2 expression is under the control of hGFAP promoter to target astrocytes, which can be reprogrammed into DCX+ neuroblasts within 4 weeks after virus injection into the spinal cord. These reprogrammed neuroblasts can further differentiate into mature neurons that express NeuN and Map 2 by the treatment of a histone deacetylase inhibitor valproic acid (VPA). This is the first report in the adult spinal cord on in vivo reprogramming astrocytes into mature neurons. However, the neuronal reprogramming efficiency by Sox2 is quite low (6% with the treatment of VPA). Whether this low number of newborn neurons will have a significant impact in functional repair after spinal cord injury is still a question. The second report from the same lab demonstrated that the p53 pathway regulates Sox2-mediated reprogramming, and suppressing p53 and p21 expression promoted cell cycle exit of the newborn neuroblasts [93]. In the presence of exogenous neurotrophic factors, these post-mitotic neurons were able to mature and acquire markers of neuronal subtypes in the adult spinal cord. Nevertheless, further investigation is needed to examine electrophysiological properties of these newborn neurons and their functional effects in behavioral assays. Another important issue to be addressed is the survival of the reprogrammed neurons in an environment suffered from spinal cord injury, because long-term survival of the reprogrammed neurons is essential for functional recovery. At least some of the Sox2-reprogrammed neurons were able to survive for up to 30 weeks in the spinal cord [92]. The critical data much needed for future studies is the regeneration of a large number of new neurons in a severe spinal cord injury model with a clear functional repair outcome.

10.10 Retina

In fish, but not mammals, Muller glia in the retina have the capacity to regenerate neurons for repair after injury. The question is whether unique mechanisms underlying neuronal regeneration in fish can be applied in mammals for functional repair. Inspired by this idea, two groups reported recently that Muller glia indeed can be reprogrammed to give rise to functional neurons that integrate into the local neuronal circuitry and even recover visual function [168, 169]. Overexpressing neurogenic transcription factor Ascl1 in Muller glia of the adult mouse retina, when combined with the histone deacetylase inhibitor trichostatin A (TSA), can successfully convert the adult glial cells into functional neurons in an excitotoxicity injury model [168]. Interestingly, the key of this successful neuronal conversion is to treat adult mice with TSA, which modifies chromatin structure for active gene transcription. The same research group has previously demonstrated that neuronal conversion by Ascl1 overexpression alone in the retina can only occur in young mice, but not adult mice, without TSA treatment [170]. Ascl1 was chosen in these studies based on its expression pattern during retinal development [171] and its upregulated expression level in Muller glia upon injury [172, 173].

Using a different set of transcription factors, another group reported that Muller glia in the adult mouse retina can be reprogrammed to a specific subtype of retinal neurons, i.e., photoreceptors, which help improve visual behavior of the mice [169]. In this study, the authors first activated Muller glia by overexpressing β-catenin to stimulate their proliferation and then 2 weeks later transduce Muller glia with a combination of three transcription factors (Otx2, Crx, and Nrl), all of which are important for photoreceptor cell fate determination. The reprogrammed photoreceptors acquire typical morphology of the native ones and express essential Rod photoreceptor genes including rhodopsin and peripherin-2 [169]. More importantly, the authors showed that photoreceptor reprogramming was able to correct visual defects of $Gnat1^{rd17}Gnat2^{cpfl3}$ transgenic mice [169]. The abovementioned two recent studies demonstrated that Muller glia in the retina can be reprogrammed, not only into functional neurons but also neurons with different subtypes, i.e., photoreceptors [169] and bipolar cells [168], indicating the plasticity of Muller glia during neuronal reprogramming process. Regeneration of other neuronal subtypes such as retinal ganglion cells is in need to fully repair retinal circuitry after injury or in diseases.

10.11 Conclusion and Perspective

Classical cell therapy relies upon replacing one cell with an external cell that has been transplanted from an external source. Such cell replacement therapy has not been widely used in therapeutic applications after being tested for several decades, partly due to low efficiency differentiation and partly due to low survival rate after engraftment. Previously, the engraftment of human fetal tissue after abortion or the use of human embryonic stem cells has met challenges of ethical issues and the limitation of resources. The invention of human iPSCs not only makes it possible to obtain unlimited human cells for therapeutic need but also makes it possible to generate autologous cells from patients' own cells for cell replacement therapy. Moreover, the concept of turning a fibroblast cell into a stem cell stimulated the imagination of scientists and makes it possible now to convert almost any type of cells into another type of different cells, an unthinkable approach decades ago. On the other hand, iPSC approach relies heavily on in vitro culture condition, which greatly limits its clinical applications due to the commonly encountered problems associated with classical stem cell therapy, such as low efficiency and low survival rate. Encouraged by in vitro transdifferentiation studies, we and other groups pioneered in vivo cell conversion technology. It is rather unexpected that in vivo converting reactive astrocytes into functional neurons could reach very high efficiency, as high as 90%. It is such high in vivo cell conversion efficiency that gives neuroscientists as well as neural stem cell biologists the confidence that the in vivo neuroregeneration approach will be the next-generation cell therapy to treat neurodegenerative disorders. In fact, despite a few successful examples of cell transplantation into the brain or retina of patients, there is so far no universal cell transplantation therapy available to treat severe neurological disorders. Exploring new technology of in vivo cell conversion as a potential therapeutic treatment might open a new avenue for searching the cure of neurodegenerative diseases such as Alzheimer's disease and Parkinson's disease. On the other hand, in vivo cell conversion technology is still in its infant stage. Much more in-depth studies are needed to elucidate the molecular mechanisms underlying the conversion process and, perhaps even more importantly, to investigate the functional integration of the converted cells into preexisting circuits. If injured neural circuits can be rebuilt through in vivo glia-to-neuron conversion, functional recovery will follow the circuit rebuilding.

References

1. Osawa M, Hanada K, Hamada H, Nakauchi H. Long-term lymphohematopoietic reconstitution by a single CD34-low/negative hematopoietic stem cell. Science. 1996;273:242–5.
2. Trounson A, McDonald C. Stem cell therapies in clinical trials: progress and challenges. Cell Stem Cell. 2015;17:11–22.
3. Enzmann GU, Benton RL, Talbott JF, Cao Q, Whittemore SR. Functional considerations of stem cell transplantation therapy for spinal cord repair. J Neurotrauma. 2006;23:479–95.
4. Gu YL, Yin LW, Zhang Z, Liu J, Liu SJ, Zhang LF, Wang TH. Neurotrophin expression in neural stem cells grafted acutely to transected spinal cord of adult rats linked to functional improvement. Cell Mol Neurobiol. 2012;32:1089–97.
5. Kamei N, Tanaka N, Oishi Y, Hamasaki T, Nakanishi K, Sakai N, Ochi M. BDNF, NT-3, and NGF released from transplanted neural progenitor cells promote corticospinal axon growth in organotypic cocultures. Spine. 2007;32:1272–8.
6. Lu P, Jones LL, Snyder EY, Tuszynski MH. Neural stem cells constitutively secrete neurotrophic factors and promote extensive host axonal growth after spinal cord injury. Exp Neurol. 2003;181:115–29.
7. Li JY, Englund E, Holton JL, Soulet D, Hagell P, Lees AJ, Lashley T, Quinn NP, Rehncrona S, Bjorklund A, et al. Lewy bodies in grafted neurons in subjects with Parkinson's disease suggest host-to-graft disease propagation. Nat Med. 2008;14:501–3.
8. Southwell DG, Nicholas CR, Basbaum AI, Stryker MP, Kriegstein AR, Rubenstein JL, Alvarez-Buylla A. Interneurons from embryonic development to cell-based therapy. Science. 2014;344:1240622.
9. Tyson JA, Anderson SA. GABAergic interneuron transplants to study development and treat disease. Trends Neurosci. 2014;37:169–77.
10. Baraban SC, Southwell DG, Estrada RC, Jones DL, Sebe JY, Alfaro-Cervello C, Garcia-Verdugo JM, Rubenstein JL, Alvarez-Buylla A. Reduction of seizures by transplantation of cortical GABAergic interneuron precursors into Kv1.1 mutant mice. Proc Natl Acad Sci U S A. 2009;106:15472–7.
11. Hunt RF, Girskis KM, Rubenstein JL, Alvarez-Buylla A, Baraban SC. GABA progenitors grafted into the adult epileptic brain control seizures and abnormal behavior. Nat Neurosci. 2013;16:692–7.

12. Martinez-Cerdeno V, Noctor SC, Espinosa A, Ariza J, Parker P, Orasji S, Daadi MM, Bankiewicz K, Alvarez-Buylla A, Kriegstein AR. Embryonic MGE precursor cells grafted into adult rat striatum integrate and ameliorate motor symptoms in 6-OHDA-lesioned rats. Cell Stem Cell. 2010;6:238–50.
13. Shetty AK, Zaman V, Hattiangady B. Repair of the injured adult hippocampus through graft-mediated modulation of the plasticity of the dentate gyrus in a rat model of temporal lobe epilepsy. J Neurosci. 2005;25:8391–401.
14. Tanaka DH, Toriumi K, Kubo K, Nabeshima T, Nakajima K. GABAergic precursor transplantation into the prefrontal cortex prevents phencyclidine-induced cognitive deficits. J Neurosci. 2011;31:14116–25.
15. Braz JM, Sharif-Naeini R, Vogt D, Kriegstein A, Alvarez-Buylla A, Rubenstein JL, Basbaum AI. Forebrain GABAergic neuron precursors integrate into adult spinal cord and reduce injury-induced neuropathic pain. Neuron. 2012;74:663–75.
16. Thomson JA, Itskovitz-Eldor J, Shapiro SS, Waknitz MA, Swiergiel JJ, Marshall VS, Jones JM. Embryonic stem cell lines derived from human blastocysts. Science. 1998;282:1145–7.
17. Goulburn AL, Stanley EG, Elefanty AG, Anderson SA. Generating GABAergic cerebral cortical interneurons from mouse and human embryonic stem cells. Stem Cell Res. 2012;8:416–26.
18. Perrier AL, Tabar V, Barberi T, Rubio ME, Bruses J, Topf N, Harrison NL, Studer L. Derivation of midbrain dopamine neurons from human embryonic stem cells. Proc Natl Acad Sci U S A. 2004;101:12543–8.
19. Wada T, Honda M, Minami I, Tooi N, Amagai Y, Nakatsuji N, Aiba K. Highly efficient differentiation and enrichment of spinal motor neurons derived from human and monkey embryonic stem cells. PLoS One. 2009;4:e6722.
20. Wichterle H, Lieberam I, Porter JA, Jessell TM. Directed differentiation of embryonic stem cells into motor neurons. Cell. 2002;110:385–97.
21. Doetsch F, Caille I, Lim DA, Garcia-Verdugo JM, Alvarez-Buylla A. Subventricular zone astrocytes are neural stem cells in the adult mammalian brain. Cell. 1999;97:703–16.
22. Johansson CB, Momma S, Clarke DL, Risling M, Lendahl U, Frisen J. Identification of a neural stem cell in the adult mammalian central nervous system. Cell. 1999;96:25–34.
23. Seri B, Garcia-Verdugo JM, McEwen BS, Alvarez-Buylla A. Astrocytes give rise to new neurons in the adult mammalian hippocampus. J Neurosci. 2001;21:7153–60.
24. Gage FH. Mammalian neural stem cells. Science. 2000;287:1433–8.
25. Curtis MA, Kam M, Nannmark U, Anderson MF, Axell MZ, Wikkelso C, Holtas S, van Roon-Mom WM, Bjork-Eriksson T, Nordborg C, et al. Human neuroblasts migrate to the olfactory bulb via a lateral ventricular extension. Science. 2007;315:1243–9.
26. Eriksson PS, Perfilieva E, Bjork-Eriksson T, Alborn AM, Nordborg C, Peterson DA, Gage FH. Neurogenesis in the adult human hippocampus. Nat Med. 1998;4:1313–7.
27. Gheusi G, Cremer H, McLean H, Chazal G, Vincent JD, Lledo PM. Importance of newly generated neurons in the adult olfactory bulb for odor discrimination. Proc Natl Acad Sci U S A. 2000;97:1823–8.
28. Lois C, Alvarez-Buylla A. Long-distance neuronal migration in the adult mammalian brain. Science. 1994;264:1145–8.
29. Lois C, Garcia-Verdugo JM, Alvarez-Buylla A. Chain migration of neuronal precursors. Science. 1996;271:978–81.
30. Petreanu L, Alvarez-Buylla A. Maturation and death of adult-born olfactory bulb granule neurons: role of olfaction. J Neurosci. 2002;22:6106–13.
31. Rochefort C, Gheusi G, Vincent JD, Lledo PM. Enriched odor exposure increases the number of newborn neurons in the adult olfactory bulb and improves odor memory. J Neurosci. 2002;22:2679–89.
32. Aimone JB, Deng W, Gage FH. Resolving new memories: a critical look at the dentate gyrus, adult neurogenesis, and pattern separation. Neuron. 2011;70:589–96.
33. Sahay A, Scobie KN, Hill AS, O'Carroll CM, Kheirbek MA, Burghardt NS, Fenton AA, Dranovsky A, Hen R. Increasing adult hippocampal neurogenesis is sufficient to improve pattern separation. Nature. 2011;472:466–70.

34. Shors TJ, Miesegaes G, Beylin A, Zhao M, Rydel T, Gould E. Neurogenesis in the adult is involved in the formation of trace memories. Nature. 2001;410:372–6.
35. Kempermann G, Kuhn HG, Gage FH. More hippocampal neurons in adult mice living in an enriched environment. Nature. 1997;386:493–5.
36. Gould E, Beylin A, Tanapat P, Reeves A, Shors TJ. Learning enhances adult neurogenesis in the hippocampal formation. Nat Neurosci. 1999;2:260–5.
37. van Praag H, Kempermann G, Gage FH. Running increases cell proliferation and neurogenesis in the adult mouse dentate gyrus. Nat Neurosci. 1999;2:266–70.
38. Ming GL, Song H. Adult neurogenesis in the mammalian brain: significant answers and significant questions. Neuron. 2011;70:687–702.
39. Arvidsson A, Collin T, Kirik D, Kokaia Z, Lindvall O. Neuronal replacement from endogenous precursors in the adult brain after stroke. Nat Med. 2002;8:963–70.
40. Zhang RL, LeTourneau Y, Gregg SR, Wang Y, Toh Y, Robin AM, Zhang ZG, Chopp M. Neuroblast division during migration toward the ischemic striatum: a study of dynamic migratory and proliferative characteristics of neuroblasts from the subventricular zone. J Neurosci. 2007;27:3157–62.
41. Faulkner RL, Jang MH, Liu XB, Duan X, Sailor KA, Kim JY, Ge S, Jones EG, Ming GL, Song H, et al. Development of hippocampal mossy fiber synaptic outputs by new neurons in the adult brain. Proc Natl Acad Sci U S A. 2008;105:14157–62.
42. Toni N, Laplagne DA, Zhao C, Lombardi G, Ribak CE, Gage FH, Schinder AF. Neurons born in the adult dentate gyrus form functional synapses with target cells. Nat Neurosci. 2008;11:901–7.
43. van Praag H, Schinder AF, Christie BR, Toni N, Palmer TD, Gage FH. Functional neurogenesis in the adult hippocampus. Nature. 2002;415:1030–4.
44. Gould E. How widespread is adult neurogenesis in mammals? Nat Rev Neurosci. 2007;8:481–8.
45. Buffo A, Rite I, Tripathi P, Lepier A, Colak D, Horn AP, Mori T, Gotz M. Origin and progeny of reactive gliosis: a source of multipotent cells in the injured brain. Proc Natl Acad Sci U S A. 2008;105:3581–6.
46. Boldrini M, Fulmore CA, Tartt AN, Simeon LR, Pavlova I, Poposka V, Rosoklija GB, Stankov A, Arango V, Dwork AJ, et al. Human hippocampal neurogenesis persists throughout aging. Cell Stem Cell. 2018;22:589–99. e585.
47. Sorrells SF, Paredes MF, Cebrian-Silla A, Sandoval K, Qi D, Kelley KW, James D, Mayer S, Chang J, Auguste KI, et al. Human hippocampal neurogenesis drops sharply in children to undetectable levels in adults. Nature. 2018;555:377–81.
48. Sun D. Endogenous neurogenic cell response in the mature mammalian brain following traumatic injury. Exp Neurol. 2015;275:405–10.
49. Gurdon JB. The developmental capacity of nuclei taken from intestinal epithelium cells of feeding tadpoles. J Embryol Exp Morphol. 1962;10:622–40.
50. Wilmut I, Schnieke AE, McWhir J, Kind AJ, Campbell KH. Viable offspring derived from fetal and adult mammalian cells. Nature. 1997;385:810–3.
51. Liu Z, Cai Y, Wang Y, Nie Y, Zhang C, Xu Y, Zhang X, Lu Y, Wang Z, Poo M, et al. Cloning of macaque monkeys by somatic cell nuclear transfer. Cell. 2018;172:881–7. e887.
52. Tapscott SJ, Davis RL, Thayer MJ, Cheng PF, Weintraub H, Lassar AB. MyoD1: a nuclear phosphoprotein requiring a Myc homology region to convert fibroblasts to myoblasts. Science. 1988;242:405–11.
53. Takahashi K, Tanabe K, Ohnuki M, Narita M, Ichisaka T, Tomoda K, Yamanaka S. Induction of pluripotent stem cells from adult human fibroblasts by defined factors. Cell. 2007;131:861–72.
54. Takahashi K, Yamanaka S. Induction of pluripotent stem cells from mouse embryonic and adult fibroblast cultures by defined factors. Cell. 2006;126:663–76.
55. Feng B, Jiang J, Kraus P, Ng JH, Heng JC, Chan YS, Yaw LP, Zhang W, Loh YH, Han J, et al. Reprogramming of fibroblasts into induced pluripotent stem cells with orphan nuclear receptor Esrrb. Nat Cell Biol. 2009;11:197–203.

56. Hanna J, Markoulaki S, Schorderet P, Carey BW, Beard C, Wernig M, Creyghton MP, Steine EJ, Cassady JP, Foreman R, et al. Direct reprogramming of terminally differentiated mature B lymphocytes to pluripotency. Cell. 2008;133:250–64.
57. Mali P, Ye Z, Hommond HH, Yu X, Lin J, Chen G, Zou J, Cheng L. Improved efficiency and pace of generating induced pluripotent stem cells from human adult and fetal fibroblasts. Stem Cells. 2008;26:1998–2005.
58. Yu J, Vodyanik MA, Smuga-Otto K, Antosiewicz-Bourget J, Frane JL, Tian S, Nie J, Jonsdottir GA, Ruotti V, Stewart R, et al. Induced pluripotent stem cell lines derived from human somatic cells. Science. 2007;318:1917–20.
59. Zhao Y, Yin X, Qin H, Zhu F, Liu H, Yang W, Zhang Q, Xiang C, Hou P, Song Z, et al. Two supporting factors greatly improve the efficiency of human iPSC generation. Cell Stem Cell. 2008;3:475–9.
60. Judson RL, Babiarz JE, Venere M, Blelloch R. Embryonic stem cell-specific microRNAs promote induced pluripotency. Nat Biotechnol. 2009;27:459–61.
61. Li Z, Yang CS, Nakashima K, Rana TM. Small RNA-mediated regulation of iPS cell generation. EMBO J. 2011;30:823–34.
62. Lin SL, Chang DC, Chang-Lin S, Lin CH, Wu DT, Chen DT, Ying SY. Mir-302 reprograms human skin cancer cells into a pluripotent ES-cell-like state. RNA. 2008;14:2115–24.
63. Subramanyam D, Lamouille S, Judson RL, Liu JY, Bucay N, Derynck R, Blelloch R. Multiple targets of miR-302 and miR-372 promote reprogramming of human fibroblasts to induced pluripotent stem cells. Nat Biotechnol. 2011;29:443–8.
64. Hou P, Li Y, Zhang X, Liu C, Guan J, Li H, Zhao T, Ye J, Yang W, Liu K, et al. Pluripotent stem cells induced from mouse somatic cells by small-molecule compounds. Science. 2013;341:651–4.
65. Huangfu D, Maehr R, Guo W, Eijkelenboom A, Snitow M, Chen AE, Melton DA. Induction of pluripotent stem cells by defined factors is greatly improved by small-molecule compounds. Nat Biotechnol. 2008;26:795–7.
66. Shi Y, Desponts C, Do JT, Hahm HS, Scholer HR, Ding S. Induction of pluripotent stem cells from mouse embryonic fibroblasts by Oct4 and Klf4 with small-molecule compounds. Cell Stem Cell. 2008;3:568–74.
67. Aoi T, Yae K, Nakagawa M, Ichisaka T, Okita K, Takahashi K, Chiba T, Yamanaka S. Generation of pluripotent stem cells from adult mouse liver and stomach cells. Science. 2008;321:699–702.
68. Stadtfeld M, Brennand K, Hochedlinger K. Reprogramming of pancreatic beta cells into induced pluripotent stem cells. Curr Biol. 2008;18:890–4.
69. Eminli S, Utikal J, Arnold K, Jaenisch R, Hochedlinger K. Reprogramming of neural progenitor cells into induced pluripotent stem cells in the absence of exogenous Sox2 expression. Stem Cells. 2008;26:2467–74.
70. Seki T, Yuasa S, Oda M, Egashira T, Yae K, Kusumoto D, Nakata H, Tohyama S, Hashimoto H, Kodaira M, et al. Generation of induced pluripotent stem cells from human terminally differentiated circulating T cells. Cell Stem Cell. 2010;7:11–4.
71. Giorgetti A, Montserrat N, Aasen T, Gonzalez F, Rodriguez-Piza I, Vassena R, Raya A, Boue S, Barrero MJ, Corbella BA, et al. Generation of induced pluripotent stem cells from human cord blood using OCT4 and SOX2. Cell Stem Cell. 2009;5:353–7.
72. Araki R, Uda M, Hoki Y, Sunayama M, Nakamura M, Ando S, Sugiura M, Ideno H, Shimada A, Nifuji A, et al. Negligible immunogenicity of terminally differentiated cells derived from induced pluripotent or embryonic stem cells. Nature. 2013;494:100–4.
73. Brennand KJ, Simone A, Jou J, Gelboin-Burkhart C, Tran N, Sangar S, Li Y, Mu Y, Chen G, Yu D, et al. Modelling schizophrenia using human induced pluripotent stem cells. Nature. 2011;473:221–5.
74. Marchetto MC, Carromeu C, Acab A, Yu D, Yeo GW, Mu Y, Chen G, Gage FH, Muotri AR. A model for neural development and treatment of Rett syndrome using human induced pluripotent stem cells. Cell. 2010;143:527–39.

75. Okano H, Nakamura M, Yoshida K, Okada Y, Tsuji O, Nori S, Ikeda E, Yamanaka S, Miura K. Steps toward safe cell therapy using induced pluripotent stem cells. Circ Res. 2013;112:523–33.
76. Okita K, Nagata N, Yamanaka S. Immunogenicity of induced pluripotent stem cells. Circ Res. 2011;109:720–1.
77. Zhao T, Zhang ZN, Rong Z, Xu Y. Immunogenicity of induced pluripotent stem cells. Nature. 2011;474:212–5.
78. Lee AS, Tang C, Rao MS, Weissman IL, Wu JC. Tumorigenicity as a clinical hurdle for pluripotent stem cell therapies. Nat Med. 2013;19:998–1004.
79. Kobayashi Y, Okada Y, Itakura G, Iwai H, Nishimura S, Yasuda A, Nori S, Hikishima K, Konomi T, Fujiyoshi K, et al. Pre-evaluated safe human iPSC-derived neural stem cells promote functional recovery after spinal cord injury in common marmoset without tumorigenicity. PLoS One. 2012;7:e52787.
80. Tsuji O, Miura K, Okada Y, Fujiyoshi K, Mukaino M, Nagoshi N, Kitamura K, Kumagai G, Nishino M, Tomisato S, et al. Therapeutic potential of appropriately evaluated safe-induced pluripotent stem cells for spinal cord injury. Proc Natl Acad Sci U S A. 2010;107:12704–9.
81. Zhou Q, Brown J, Kanarek A, Rajagopal J, Melton DA. In vivo reprogramming of adult pancreatic exocrine cells to beta-cells. Nature. 2008;455:627–32.
82. Ieda M, Fu JD, Delgado-Olguin P, Vedantham V, Hayashi Y, Bruneau BG, Srivastava D. Direct reprogramming of fibroblasts into functional cardiomyocytes by defined factors. Cell. 2010;142:375–86.
83. Qian L, Huang Y, Spencer CI, Foley A, Vedantham V, Liu L, Conway SJ, Fu JD, Srivastava D. In vivo reprogramming of murine cardiac fibroblasts into induced cardiomyocytes. Nature. 2012;485:593–8.
84. Song K, Nam YJ, Luo X, Qi X, Tan W, Huang GN, Acharya A, Smith CL, Tallquist MD, Neilson EG, et al. Heart repair by reprogramming non-myocytes with cardiac transcription factors. Nature. 2012;485:599–604.
85. Brulet R, Matsuda T, Zhang L, Miranda C, Giacca M, Kaspar BK, Nakashima K, Hsieh J. NEUROD1 instructs neuronal conversion in non-reactive astrocytes. Stem Cell Reports. 2017;8:1506–15.
86. Gascon S, Murenu E, Masserdotti G, Ortega F, Russo GL, Petrik D, Deshpande A, Heinrich C, Karow M, Robertson SP, et al. Identification and successful negotiation of a metabolic checkpoint in direct neuronal reprogramming. Cell Stem Cell. 2015;18:396–409.
87. Grande A, Sumiyoshi K, Lopez-Juarez A, Howard J, Sakthivel B, Aronow B, Campbell K, Nakafuku M. Environmental impact on direct neuronal reprogramming in vivo in the adult brain. Nat Commun. 2013;4:2373.
88. Guo Z, Zhang L, Wu Z, Chen Y, Wang F, Chen G. In vivo direct reprogramming of reactive glial cells into functional neurons after brain injury and in an Alzheimer's disease model. Cell Stem Cell. 2014;14:188–202.
89. Liu Y, Miao Q, Yuan J, Han S, Zhang P, Li S, Rao Z, Zhao W, Ye Q, Geng J, et al. Ascl1 converts dorsal midbrain astrocytes into functional neurons in vivo. J Neurosci. 2015;35:9336–55.
90. Niu W, Zang T, Smith DK, Vue TY, Zou Y, Bachoo R, Johnson JE, Zhang CL. SOX2 reprograms resident astrocytes into neural progenitors in the adult brain. Stem Cell Reports. 2015;4:780–94.
91. Niu W, Zang T, Zou Y, Fang S, Smith DK, Bachoo R, Zhang CL. In vivo reprogramming of astrocytes to neuroblasts in the adult brain. Nat Cell Biol. 2013;15:1164–75.
92. Su Z, Niu W, Liu ML, Zou Y, Zhang CL. In vivo conversion of astrocytes to neurons in the injured adult spinal cord. Nat Commun. 2014;5:3338.
93. Wang LL, Su Z, Tai W, Zou Y, Xu XM, Zhang CL. The p53 pathway controls SOX2-mediated reprogramming in the adult mouse spinal cord. Cell Rep. 2016;17:891–903.
94. Rezvani M, Espanol-Suner R, Malato Y, Dumont L, Grimm AA, Kienle E, Bindman JG, Wiedtke E, Hsu BY, Naqvi SJ, et al. In vivo hepatic reprogramming of myofibroblasts with AAV vectors as a therapeutic strategy for liver fibrosis. Cell Stem Cell. 2016;18:809–16.

95. Song G, Pacher M, Balakrishnan A, Yuan Q, Tsay HC, Yang D, Reetz J, Brandes S, Dai Z, Putzer BM, et al. Direct reprogramming of hepatic myofibroblasts into hepatocytes in vivo attenuates liver fibrosis. Cell Stem Cell. 2016;18:797–808.
96. Ionescu-Tirgoviste C, Gagniuc PA, Gubceac E, Mardare L, Popescu I, Dima S, Militaru M. A 3D map of the islet routes throughout the healthy human pancreas. Sci Rep. 2015;5:14634.
97. Rutter GA. GABA signaling: a route to new pancreatic beta cells. Cell Res. 2017;27:309–10.
98. Bramswig NC, Everett LJ, Schug J, Dorrell C, Liu C, Luo Y, Streeter PR, Naji A, Grompe M, Kaestner KH. Epigenomic plasticity enables human pancreatic alpha to beta cell reprogramming. J Clin Invest. 2013;123:1275–84.
99. Collombat P, Xu X, Ravassard P, Sosa-Pineda B, Dussaud S, Billestrup N, Madsen OD, Serup P, Heimberg H, Mansouri A. The ectopic expression of Pax4 in the mouse pancreas converts progenitor cells into alpha and subsequently beta cells. Cell. 2009;138:449–62.
100. Courtney M, Gjernes E, Druelle N, Ravaud C, Vieira A, Ben-Othman N, Pfeifer A, Avolio F, Leuckx G, Lacas-Gervais S, et al. The inactivation of Arx in pancreatic alpha-cells triggers their neogenesis and conversion into functional beta-like cells. PLoS Genet. 2013;9:e1003934.
101. Cheng CW, Villani V, Buono R, Wei M, Kumar S, Yilmaz OH, Cohen P, Sneddon JB, Perin L, Longo VD. Fasting-mimicking diet promotes Ngn3-driven beta-cell regeneration to reverse diabetes. Cell. 2017;168:775–88. e712.
102. Ben-Othman N, Vieira A, Courtney M, Record F, Gjernes E, Avolio F, Hadzic B, Druelle N, Napolitano T, Navarro-Sanz S, et al. Long-term GABA administration induces alpha cell-mediated beta-like cell neogenesis. Cell. 2017;168:73–85.e11.
103. Li J, Casteels T, Frogne T, Ingvorsen C, Honore C, Courtney M, Huber KV, Schmitner N, Kimmel RA, Romanov RA, et al. Artemisinins target GABAA receptor signaling and impair alpha cell identity. Cell. 2017;168:86–100. e115.
104. Beltrami AP, Barlucchi L, Torella D, Baker M, Limana F, Chimenti S, Kasahara H, Rota M, Musso E, Urbanek K, et al. Adult cardiac stem cells are multipotent and support myocardial regeneration. Cell. 2003;114:763–76.
105. Behfar A, Crespo-Diaz R, Terzic A, Gersh BJ. Cell therapy for cardiac repair-lessons from clinical trials. Nat Rev Cardiol. 2014;11:232–46.
106. Yang X, Pabon L, Murry CE. Engineering adolescence: maturation of human pluripotent stem cell-derived cardiomyocytes. Circ Res. 2014;114:511–23.
107. Carpenter L, Carr C, Yang CT, Stuckey DJ, Clarke K, Watt SM. Efficient differentiation of human induced pluripotent stem cells generates cardiac cells that provide protection following myocardial infarction in the rat. Stem Cells Dev. 2012;21:977–86.
108. Li X, Zhang F, Song G, Gu W, Chen M, Yang B, Li D, Wang D, Cao K. Intramyocardial injection of pig pluripotent stem cells improves left ventricular function and perfusion: a study in a porcine model of acute myocardial infarction. PLoS One. 2013;8:e66688.
109. Mauritz C, Martens A, Rojas SV, Schnick T, Rathert C, Schecker N, Menke S, Glage S, Zweigerdt R, Haverich A, et al. Induced pluripotent stem cell (iPSC)-derived Flk-1 progenitor cells engraft, differentiate, and improve heart function in a mouse model of acute myocardial infarction. Eur Heart J. 2011;32:2634–41.
110. Nelson TJ, Martinez-Fernandez A, Yamada S, Perez-Terzic C, Ikeda Y, Terzic A. Repair of acute myocardial infarction by human stemness factors induced pluripotent stem cells. Circulation. 2009;120:408–16.
111. David R, Brenner C, Stieber J, Schwarz F, Brunner S, Vollmer M, Mentele E, Muller-Hocker J, Kitajima S, Lickert H, et al. MesP1 drives vertebrate cardiovascular differentiation through Dkk-1-mediated blockade of Wnt-signalling. Nat Cell Biol. 2008;10:338–45.
112. Murry CE, Kay MA, Bartosek T, Hauschka SD, Schwartz SM. Muscle differentiation during repair of myocardial necrosis in rats via gene transfer with MyoD. J Clin Invest. 1996;98:2209–17.
113. Reiter JF, Alexander J, Rodaway A, Yelon D, Patient R, Holder N, Stainier DY. Gata5 is required for the development of the heart and endoderm in zebrafish. Genes Dev. 1999;13:2983–95.

114. Takeuchi JK, Bruneau BG. Directed transdifferentiation of mouse mesoderm to heart tissue by defined factors. Nature. 2009;459:708–11.
115. Inagawa K, Miyamoto K, Yamakawa H, Muraoka N, Sadahiro T, Umei T, Wada R, Katsumata Y, Kaneda R, Nakade K, et al. Induction of cardiomyocyte-like cells in infarct hearts by gene transfer of Gata4, Mef2c, and Tbx5. Circ Res. 2012;111:1147–56.
116. Jayawardena TM, Egemnazarov B, Finch EA, Zhang L, Payne JA, Pandya K, Zhang Z, Rosenberg P, Mirotsou M, Dzau VJ. MicroRNA-mediated in vitro and in vivo direct reprogramming of cardiac fibroblasts to cardiomyocytes. Circ Res. 2012;110:1465–73.
117. Jayawardena TM, Finch EA, Zhang L, Zhang H, Hodgkinson CP, Pratt RE, Rosenberg PB, Mirotsou M, Dzau VJ. MicroRNA induced cardiac reprogramming in vivo: evidence for mature cardiac myocytes and improved cardiac function. Circ Res. 2015;116:418–24.
118. Addis RC, Ifkovits JL, Pinto F, Kellam LD, Esteso P, Rentschler S, Christoforou N, Epstein JA, Gearhart JD. Optimization of direct fibroblast reprogramming to cardiomyocytes using calcium activity as a functional measure of success. J Mol Cell Cardiol. 2013;60:97–106.
119. Wang H, Cao N, Spencer CI, Nie B, Ma T, Xu T, Zhang Y, Wang X, Srivastava D, Ding S. Small molecules enable cardiac reprogramming of mouse fibroblasts with a single factor, Oct4. Cell Rep. 2014;6:951–60.
120. Muraoka N, Yamakawa H, Miyamoto K, Sadahiro T, Umei T, Isomi M, Nakashima H, Akiyama M, Wada R, Inagawa K, et al. MiR-133 promotes cardiac reprogramming by directly repressing Snai1 and silencing fibroblast signatures. EMBO J. 2014;33:1565–81.
121. Ifkovits JL, Addis RC, Epstein JA, Gearhart JD. Inhibition of TGFbeta signaling increases direct conversion of fibroblasts to induced cardiomyocytes. PLoS One. 2014;9:e89678.
122. Mohamed TM, Stone NR, Berry EC, Radzinsky E, Huang Y, Pratt K, Ang YS, Yu P, Wang H, Tang S, et al. Chemical enhancement of in vitro and in vivo direct cardiac reprogramming. Circulation. 2017;135:978–95.
123. Miyamoto K, Akiyama M, Tamura F, Isomi M, Yamakawa H, Sadahiro T, Muraoka N, Kojima H, Haginiwa S, Kurotsu S, et al. Direct in vivo reprogramming with Sendai virus vectors improves cardiac function after myocardial infarction. Cell Stem Cell. 2018;22:91–103. e105.
124. Zhao Y, Londono P, Cao Y, Sharpe EJ, Proenza C, O'Rourke R, Jones KL, Jeong MY, Walker LA, Buttrick PM, et al. High-efficiency reprogramming of fibroblasts into cardiomyocytes requires suppression of pro-fibrotic signalling. Nat Commun. 2015;6:8243.
125. Zhou Y, Wang L, Vaseghi HR, Liu Z, Lu R, Alimohamadi S, Yin C, Fu JD, Wang GG, Liu J, et al. Bmi1 is a key epigenetic barrier to direct cardiac reprogramming. Cell Stem Cell. 2016;18:382–95.
126. Quinones MJ, Leor J, Kloner RA, Ito M, Patterson M, Witke WF, Kedes L. Avoidance of immune response prolongs expression of genes delivered to the adult rat myocardium by replication-defective adenovirus. Circulation. 1996;94:1394–401.
127. Ebrahimi B. In vivo reprogramming for heart regeneration: a glance at efficiency, environmental impacts, challenges and future directions. J Mol Cell Cardiol. 2017;108:61–72.
128. Mathison M, Gersch RP, Nasser A, Lilo S, Korman M, Fourman M, Hackett N, Shroyer K, Yang J, Ma Y, et al. In vivo cardiac cellular reprogramming efficacy is enhanced by angiogenic preconditioning of the infarcted myocardium with vascular endothelial growth factor. J Am Heart Assoc. 2012;1:e005652.
129. Yamakawa H, Muraoka N, Miyamoto K, Sadahiro T, Isomi M, Haginiwa S, Kojima H, Umei T, Akiyama M, Kuishi Y, et al. Fibroblast growth factors and vascular endothelial growth factor promote cardiac reprogramming under defined conditions. Stem Cell Reports. 2015;5:1128–42.
130. Islas JF, Liu Y, Weng KC, Robertson MJ, Zhang S, Prejusa A, Harger J, Tikhomirova D, Chopra M, Iyer D, et al. Transcription factors ETS2 and MESP1 transdifferentiate human dermal fibroblasts into cardiac progenitors. Proc Natl Acad Sci U S A. 2012;109:13016–21.
131. Wada R, Muraoka N, Inagawa K, Yamakawa H, Miyamoto K, Sadahiro T, Umei T, Kaneda R, Suzuki T, Kamiya K, et al. Induction of human cardiomyocyte-like cells from fibroblasts by defined factors. Proc Natl Acad Sci U S A. 2013;110:12667–72.

132. Zhang Y, Chu Y, Shen W, Dou Z. Effect of 5-azacytidine induction duration on differentiation of human first-trimester fetal mesenchymal stem cells towards cardiomyocyte-like cells. Interact Cardiovasc Thorac Surg. 2009;9:943–6.
133. Fu JD, Stone NR, Liu L, Spencer CI, Qian L, Hayashi Y, Delgado-Olguin P, Ding S, Bruneau BG, Srivastava D. Direct reprogramming of human fibroblasts toward a cardiomyocyte-like state. Stem Cell Reports. 2013;1:235–47.
134. Li XH, Li Q, Jiang L, Deng C, Liu Z, Fu Y, Zhang M, Tan H, Feng Y, Shan Z, et al. Generation of functional human cardiac progenitor cells by high-efficiency protein transduction. Stem Cells Transl Med. 2015;4:1415–24.
135. Cao N, Huang Y, Zheng J, Spencer CI, Zhang Y, Fu JD, Nie B, Xie M, Zhang M, Wang H, et al. Conversion of human fibroblasts into functional cardiomyocytes by small molecules. Science. 2016;352:1216–20.
136. Michalopoulos GK. Liver regeneration. J Cell Physiol. 2007;213:286–300.
137. Guan Y, Xu D, Peltz G. Treating liver fibrosis: (re)programmed to succeed. Cell Stem Cell. 2016;18:683–4.
138. Faiz M, Sachewsky N, Gascon S, Bang KW, Morshead CM, Nagy A. Adult neural stem cells from the subventricular zone give rise to reactive astrocytes in the cortex after stroke. Cell Stem Cell. 2015;17:624–34.
139. Buffo A, Vosko MR, Erturk D, Hamann GF, Jucker M, Rowitch D, Gotz M. Expression pattern of the transcription factor Olig2 in response to brain injuries: implications for neuronal repair. Proc Natl Acad Sci U S A. 2005;102:18183–8.
140. Heinrich C, Blum R, Gascon S, Masserdotti G, Tripathi P, Sanchez R, Tiedt S, Schroeder T, Gotz M, Berninger B. Directing astroglia from the cerebral cortex into subtype specific functional neurons. PLoS Biol. 2010;8:e1000373.
141. Kuwabara T, Hsieh J, Muotri A, Yeo G, Warashina M, Lie DC, Moore L, Nakashima K, Asashima M, Gage FH. Wnt-mediated activation of NeuroD1 and retro-elements during adult neurogenesis. Nat Neurosci. 2009;12:1097–105.
142. Gao Z, Ure K, Ables JL, Lagace DC, Nave KA, Goebbels S, Eisch AJ, Hsieh J. Neurod1 is essential for the survival and maturation of adult-born neurons. Nat Neurosci. 2009;12:1090–2.
143. Chai H, Diaz-Castro B, Shigetomi E, Monte E, Octeau JC, Yu X, Cohn W, Rajendran PS, Vondriska TM, Whitelegge JP, et al. Neural circuit-specialized astrocytes: transcriptomic, proteomic, morphological, and functional evidence. Neuron. 2017;95:531–49. e539.
144. Hughes EG, Kang SH, Fukaya M, Bergles DE. Oligodendrocyte progenitors balance growth with self-repulsion to achieve homeostasis in the adult brain. Nat Neurosci. 2013;16:668–76.
145. Zhou H, Liu J, Zhou C, Gao N, Rao Z, Li H, Hu X, Li C, Yao X, Shen X, et al. In vivo simultaneous transcriptional activation of multiple genes in the brain using CRISPR-dCas9-activator transgenic mice. Nat Neurosci. 2018;21(3):440–6.
146. Guo JJ, Liu F, Sun X, Huang JJ, Xu M, Sun FY. Bcl-2 enhances the formation of newborn striatal long-projection neurons in adult rat brain after a transient ischemic stroke. Neurosci Bull. 2012;28:669–79.
147. Hou SW, Wang YQ, Xu M, Shen DH, Wang JJ, Huang F, Yu Z, Sun FY. Functional integration of newly generated neurons into striatum after cerebral ischemia in the adult rat brain. Stroke. 2008;39:2837–44.
148. Li L, Harms KM, Ventura PB, Lagace DC, Eisch AJ, Cunningham LA. Focal cerebral ischemia induces a multilineage cytogenic response from adult subventricular zone that is predominantly gliogenic. Glia. 2010;58:1610–9.
149. Wang YQ, Guo X, Qiu MH, Feng XY, Sun FY. VEGF overexpression enhances striatal neurogenesis in brain of adult rat after a transient middle cerebral artery occlusion. J Neurosci Res. 2007;85:73–82.
150. Yamashita T, Ninomiya M, Hernandez Acosta P, Garcia-Verdugo JM, Sunabori T, Sakaguchi M, Adachi K, Kojima T, Hirota Y, Kawase T, et al. Subventricular zone-derived neuroblasts migrate and differentiate into mature neurons in the post-stroke adult striatum. J Neurosci. 2006;26:6627–36.

151. Zhang RL, Chopp M, Roberts C, Liu X, Wei M, Nejad-Davarani SP, Wang X, Zhang ZG. Stroke increases neural stem cells and angiogenesis in the neurogenic niche of the adult mouse. PLoS One. 2014;9:e113972.
152. Magnusson JP, Goritz C, Tatarishvili J, Dias DO, Smith EM, Lindvall O, Kokaia Z, Frisen J. A latent neurogenic program in astrocytes regulated by Notch signaling in the mouse. Science. 2014;346:237–41.
153. Ghasemi-Kasman M, Hajikaram M, Baharvand H, Javan M. MicroRNA-mediated in vitro and in vivo direct conversion of astrocytes to neuroblasts. PLoS One. 2015;10:e0127878.
154. Torper O, Pfisterer U, Wolf DA, Pereira M, Lau S, Jakobsson J, Bjorklund A, Grealish S, Parmar M. Generation of induced neurons via direct conversion in vivo. Proc Natl Acad Sci U S A. 2013;110:7038–43.
155. Heinrich C, Bergami M, Gascon S, Lepier A, Vigano F, Dimou L, Sutor B, Berninger B, Gotz M. Sox2-mediated conversion of NG2 glia into induced neurons in the injured adult cerebral cortex. Stem Cell Reports. 2014;3:1000–14.
156. Rivetti di Val Cervo P, Romanov RA, Spigolon G, Masini D, Martin-Montanez E, Toledo EM, La Manno G, Feyder M, Pifl C, Ng YH, et al. Induction of functional dopamine neurons from human astrocytes in vitro and mouse astrocytes in a Parkinson's disease model. Nat Biotechnol. 2017;35:444–52.
157. Yoo J, Lee E, Kim HY, Youn DH, Jung J, Kim H, Chang Y, Lee W, Shin J, Baek S, et al. Electromagnetized gold nanoparticles mediate direct lineage reprogramming into induced dopamine neurons in vivo for Parkinson's disease therapy. Nat Nanotechnol. 2017;12(10):1006–14.
158. Ben Haim L, Rowitch DH. Functional diversity of astrocytes in neural circuit regulation. Nat Rev Neurosci. 2017;18:31–41.
159. Pereira M, Birtele M, Shrigley S, Benitez JA, Hedlund E, Parmar M, Ottosson DR. Direct reprogramming of resident NG2 glia into neurons with properties of fast-spiking parvalbumin-containing interneurons. Stem Cell Reports. 2017;9:742–51.
160. Torper O, Ottosson DR, Pereira M, Lau S, Cardoso T, Grealish S, Parmar M. In vivo reprogramming of striatal NG2 glia into functional neurons that integrate into local host circuitry. Cell Rep. 2015;12:474–81.
161. Horky LL, Galimi F, Gage FH, Horner PJ. Fate of endogenous stem/progenitor cells following spinal cord injury. J Comp Neurol. 2006;498:525–38.
162. Horner PJ, Power AE, Kempermann G, Kuhn HG, Palmer TD, Winkler J, Thal LJ, Gage FH. Proliferation and differentiation of progenitor cells throughout the intact adult rat spinal cord. J Neurosci. 2000;20:2218–28.
163. Ohori Y, Yamamoto S, Nagao M, Sugimori M, Yamamoto N, Nakamura K, Nakafuku M. Growth factor treatment and genetic manipulation stimulate neurogenesis and oligodendrogenesis by endogenous neural progenitors in the injured adult spinal cord. J Neurosci. 2006;26:11948–60.
164. Kehl LJ, Fairbanks CA, Laughlin TM, Wilcox GL. Neurogenesis in postnatal rat spinal cord: a study in primary culture. Science. 1997;276:586–9.
165. Weiss S, Dunne C, Hewson J, Wohl C, Wheatley M, Peterson AC, Reynolds BA. Multipotent CNS stem cells are present in the adult mammalian spinal cord and ventricular neuroaxis. J Neurosci. 1996;16:7599–609.
166. Onifer SM, Cannon AB, Whittemore SR. Altered differentiation of CNS neural progenitor cells after transplantation into the injured adult rat spinal cord. Cell Transplant. 1997;6:327–38.
167. Whittemore SR. Neuronal replacement strategies for spinal cord injury. J Neurotrauma. 1999;16:667–73.
168. Jorstad NL, Wilken MS, Grimes WN, Wohl SG, VandenBosch LS, Yoshimatsu T, Wong RO, Rieke F, Reh TA. Stimulation of functional neuronal regeneration from Muller glia in adult mice. Nature. 2017;548:103–7.
169. Yao K, Qiu S, Wang YV, Park SJH, Mohns EJ, Mehta B, Liu X, Chang B, Zenisek D, Crair MC, et al. Restoration of vision after de novo genesis of rod photoreceptors in mammalian retinas. Nature. 2018;560:484–8.

170. Ueki Y, Wilken MS, Cox KE, Chipman L, Jorstad N, Sternhagen K, Simic M, Ullom K, Nakafuku M, Reh TA. Transgenic expression of the proneural transcription factor Ascl1 in Muller glia stimulates retinal regeneration in young mice. Proc Natl Acad Sci U S A. 2015;112:13717–22.
171. Brzezinski JAt, Kim EJ, Johnson JE, Reh TA. Ascl1 expression defines a subpopulation of lineage-restricted progenitors in the mammalian retina. Development. 2011;138:3519–31.
172. Fausett BV, Gumerson JD, Goldman D. The proneural basic helix-loop-helix gene ascl1a is required for retina regeneration. J Neurosci. 2008;28:1109–17.
173. Fischer AJ, Reh TA. Muller glia are a potential source of neural regeneration in the postnatal chicken retina. Nat Neurosci. 2001;4:247–52.

The manufacturer's authorised representative in the EU is Springer Nature Customer Service Centre GmbH, Europaplatz 3, 69115 Heidelberg, Germany. If you have any concerns regarding our products, please contact ProductSafety@springernature.com

Printed and bound by CPI Group (UK) Ltd, Croydon, CR0 4YY
25/09/2026
02222841-0016